Lecture Notes in Physics

New Series m: Monographs

The Editorial Policy for Monographs

The series Lecture Notes in Physics reports new developments in physical research and teaching - quickly, informally, and at a high level. The type of material considered for publication in the New Series m includes monographs presenting original research or new angles in a classical field. The timeliness of a manuscript is more important than its form, which may be preliminary or tentative. Manuscripts should be reasonably self-contained. They will often present not only results of the author(s) but also related work by other people and will provide sufficient motivation, examples, and applications.

The manuscripts or a detailed description thereof should be submitted either to one of the series editors or to the managing editor. The proposal is then carefully refereed. A final decision concerning publication can often only be made on the basis of the complete manuscript, but otherwise the editors will try to make a preliminary decision as definite as they can on the basis of the available information.

Manuscripts should be no less than 100 and preferably no more than 400 pages in length. Final manuscripts should preferably be in English, or possibly in French or German. They should include a table of contents and an informative introduction accessible also to readers not particularly familiar with the topic treated. Authors are free to use the material in other publications. However, if extensive use is made elsewhere, the publisher should be informed.

Authors receive jointly 50 complimentary copies of their book. They are entitled to purchase further copies of their book at a reduced rate. As a rule no reprints of individual contributions can be supplied. No royalty is paid on Lecture Notes in Physics volumes. Commitment to publish is made by letter of interest rather than by signing a formal contract. Springer-Verlag secures the copyright for each volume.

The Production Process

The books are hardbound, and quality paper appropriate to the needs of the author(s) is used. Publication time is about ten weeks. More than twenty years of experience guarantee authors the best possible service. To reach the goal of rapid publication at a low price the technique of photographic reproduction from a camera-ready manuscript was chosen. This process shifts the main responsibility for the technical quality considerably from the publisher to the author. We therefore urge all authors to observe very carefully our guidelines for the preparation of camera-ready manuscripts, which we will supply on request. This applies especially to the quality of figures and halftones submitted for publication. Figures should be submitted as originals or glossy prints, as very often Xerox copies are not suitable for reproduction. In addition, it might be useful to look at some of the volumes already published or, especially if some atypical text is planned, to write to the Physics Editorial Department of Springer-Verlag direct. This avoids mistakes and time-consuming correspondence during the production period.

As a special service, we offer free of charge LATEX and TEX macro packages to format the text according to Springer-Verlag's quality requirements. We strongly recommend authors to make use of this offer, as the result will be a book of considerably improved technical quality. The typescript will be reduced in size (75% of the original). Therefore, e. g. any writing within figures should not be smaller than 2.5 mm.

Manuscripts not meeting the technical standard of the series will have to be returned for improvement.

For further information please contact Springer-Verlag, Physics Editorial Department II, Tiergartenstrasse 17, W-6900 Heidelberg, FRG.

Paul Busch   Pekka J. Lahti   Peter Mittelstaedt

The Quantum Theory of Measurement

Springer-Verlag

Berlin Heidelberg New York
London Paris Tokyo
Hong Kong Barcelona
Budapest

Authors

Paul Busch
Peter Mittelstaedt
Institut für Theoretische Physik, Universität zu Köln
Zülpicher Straße 77, W-5000 Köln 41, FRG

Pekka J. Lahti
Department of Physics, University of Turku
Finland

ISBN 3-540-54334-1 Springer-Verlag Berlin Heidelberg New York
ISBN 0-387-54334-1 Springer-Verlag New York Berlin Heidelberg

© Springer-Verlag Berlin Heidelberg 1991
Printed in Germany

Printing and binding: Druckhaus Beltz, Hemsbach/Bergstr.
2153/3140-543210 - Printed on acid-free paper

Acknowledgements

This work is the product of a long period of collaboration between its authors and has benefited from many sources of inspiration, as well as moral, and, not least, material, support. First of all, we are indebted to Sławek Bugajski for bringing together the German and Finnish partners, thus enabling them to experience a very rewarding time of more than ten years of joint research effort. The formation of our views on the quantum theory of measurement laid down in the present text was influenced by numerous fruitful discussions particularly with Enrico Beltrametti, Gianni Cassinelli, Bas van Fraassen, and Franklin Schroeck. Sławek Bugajski also was among the first to undertake the tedious job of critically reading and commenting on our manuscript. Our younger colleagues Bernd Fischer, Almut Kirchner, Ralf Quadt and Harald Scherer served as test readers too, and their questions and comments in seminars on the subject helped to improve the presentation of the material.

Technical assistance was provided by George Maude, lecturer in English at the University of Turku, who gave language advice and proofread the whole manuscript, and by Tanja Bieker, Harri Pietilä and Harald Scherer whose TEX-expertise and patience was of great help to us.

Finally, it is a pleasure to acknowledge financial support extended to us by the following institutions: the Academy of Finland, the Alexander von Humboldt Foundation, the Bundesministerium für Forschung und Technologie, and the University of Turku Foundation.

Preface

The present treatise is concerned with the quantum mechanical theory of measurement. Since the development of quantum theory in the 1920s the measuring process has been considered a very important problem. A large number of articles have accordingly been devoted to this subject. In this way the quantum mechanical measurement problem has been a source of inspiration for physical, mathematical and philosophical investigations into the foundations of quantum theory, which has had an impact on a great variety of research fields, ranging from the physics of macroscopic systems to probability theory and algebra. Moreover, while many steps forward have been made and much insight has been gained on the road towards a solution of the measurement problem, left open nonetheless are important questions, which have induced several interesting developments. Hence even today it cannot be said that the measurement process has lost its topicality and excitement. Moreover, research in this field has made contact with current advances in high technology, which provide new possibilities for performing former Gedanken experiments. For these reasons we felt that the time had come to develop a systematic exposition of the quantum theory of measurement which might serve as a basis and reference for future research into the foundations of quantum mechanics. But there are other sources of motivation which led us to make this effort.

First of all, in spite of the many contributions to measurement theory there is still no generally accepted approach. Much worse, a considerable fraction of even recent publications on the subject is based on an erroneous or insufficient understanding of the measurement problem. It therefore seems desirable to formulate a precise definition of the subject of quantum measurement theory. This should give rise to a systematic account of the options for solving the problem of measurement and allow for an evaluation of the various approaches. In this sense the present work may be taken as a first step towards a textbook on the quantum theory of measurement, the lack of which has been pointed out by Wheeler and Zurek (1983).

In view of the difficulties encountered in the quantum theory of measurement many distinguished authors have considered the possibility that quantum mechanics is not a universally valid theory. In particular, the question has been raised whether macroscopic systems, such as measuring devices, are beyond the scope of this theory. Adopting this point of view would allow one to reformulate, and possibly solve, the open problems of quantum mechanics within the framework of more general theories. Such far-reaching conclusions should, however, be substantiated by means of a close chain of arguments. We shall try to spell out some of the arguments that endeavour to prove the limitations of quantum mechanics in the context of measurement theory. The resulting no-go theorems naturally entail a specification of the various modifications of quantum mechanics which might lead to a satisfactory resolution. At the same time they contribute to an understanding of those interpretations maintaining the universal validity of quantum mechanics.

Next, we are not aware of the existence of a review of the measurement problem which takes into account the developments in the foundations of quantum mechanics over the past two decades. The operational language based on the notions of effects and operations, and the ensuing general concepts of observables and instruments have proved extremely useful not only in foundational issues (as documented in the monographs of Ludwig (1983a,1987), Kraus (1983), or Prugovečki (1986)), but also in applications of quantum physics in areas like quantum optics or signal processing (as represented by the books of Davies (1976), Helstrom (1973), or Holevo (1982)). These concepts must be regarded as the contemporary standards for the rigorous formulation of physical problems. They will be employed here for the precise definitions of operational and probabilistic concepts needed for uniquely fixing the notion of measurement in quantum mechanics and developing a formulation of the quantum theory of measurement general enough to cover the present scope of applications.

The introduction of general observables has shed new light on the problem of macroscopic quantum systems and the question of the (quasi-) classical limit of quantum mechanics, thus providing a redefinition of the notion of macroscopic observables. In this way a new approach to the measurement problem – unsharp objectification – has emerged in the last few years and will be sketched out in the course of our review.

The failure of the quantum theory of measurement in its original form has led several authors to propose a modified conception of dynamics, incorporating stochastic elements into the Schrödinger equation or taking into account the influence of the environment of a quantum system. In both cases the measuring process can no longer be described in terms of a unitary dynamical group. Hence the traditional theory of measurement should also be extended to cover nonunitary state transformations.

The preceding remarks suggest that the incorporation of general observables and nonunitary dynamics into quantum measurement theory necessitates, and makes possible, an entirely new approach to this theory. We shall try to bring into a systematic order the new results obtained in the course of many detailed investigations, recovering the known results as special cases. In this way we shall hope to have established a systematic description of the quantum mechanical measurement process together with a concise formulation of the measurement problem. In our view the generalized mathematical and conceptual framework of quantum mechanics referred to above allows for the first time for a proper formulation of many aspects of the measurement problem *within* this theory, thereby opening up new options for its solution. Thus it has become evident that these questions, which were sometimes considered to belong to the realm of philosophical contemplation, have assumed the status of well-defined and tractable *physical* problems.

Cologne, June 1991

Paul Busch
Pekka Lahti
Peter Mittelstaedt

Table of Contents

Chapter I

Chapter II

Chapter III

Chapter IV

Chapter V

Chapter I

General Introduction

1. The problem of measurement in quantum mechanics

An understanding of quantum mechanics in the sense of a generally accepted interpretation has not yet been attained. The ultimate reason for this difficulty must be seen in the irreducibly probabilistic structure of quantum mechanics which is rooted in the nonclassical character of its language. An operational analysis of the peculiarities of quantum mechanics shows that the interpretational problems are closely related to the difficulties of the quantum theory of measurement. It is the purpose of this review to spell out in detail these connections.

The task of the quantum theory of measurement is to investigate the semantical consistency of quantum mechanics. Phrased in general terms, quantum mechanics, as a physical theory, and the quantum theory of measurement as a part of it, are based on a "splitting" of the empirical world into four "parts": (1) object systems S (to be observed), (2) apparatus A (preparation and registration devices), (3) environments $\mathcal{E}$ (the "rest" of the physical world which one intends to ignore), and (4) observers $\mathcal{O}$. Depending on the type of interpretation in question, observers or environments may or may not be ignored in the description of the measuring process within the quantum theory of measurement. Providing that quantum mechanics is considered as a theory of individual objects, the most important questions to be answered by measurement theory are: (1) how it is possible for objects to be prepared, that is, isolated from their environments and brought into well defined states; (2) how the measurement of a given observable is achieved; and (3), how objects can be reestablished after measurements, that is, be separated from the apparatus. The underlying common problem is the *objectification problem*; that is, the question of how definite measurement outcomes are obtained.

We shall try to elucidate the status and the precise form of these questions. In Chapter II basic features of quantum mechanics are summarized which may be regarded as the root of the objectification problem. Chapter III is devoted to a systematic exposition of the quantum theory of measurement. Various solutions to the measurement problem proposed within a number of current interpretations of quantum mechanics will be reviewed in Chapter IV. Chapter V closes the treatise with our general conclusions.

In the present chapter a decision tree will be formulated as a guide to a systematic evaluation of the various interpretations of quantum mechanics. A brief historical overview of these interpretations may serve as a first orientation, showing, in passing, the origins of the present approach.

2. Historical account: interpretations and reconstructions of quantum mechanics

One may distinguish four or five overlapping phases in the development of research in the foundations of quantum mechanics.

Early discussions among the pioneers (1927-1935) led to the well-known versions of the so-called Copenhagen interpretation. In the discussions between Bohr and Heisenberg [23,96] and Bohr and Einstein (during the years 1930 - 1935; cf. Ref. 26) the quantum theory of measurement was touched upon only in an informal way. It is only in the monographs of von Neumann (1932) and Pauli (1933) that one finds the first rigorous and explicit formulations of measurement problems in the manner in which they are the subject of the present treatise.

Reconsiderations of interpretational questions extending essentially from the 1950s to the 1970s were mainly motivated by attempts to explore the possibilities of establishing *realistic* interpretations of quantum mechanics considered as a *universally valid* theory. Much of this was anticipated in and taken up from the early works of von Neumann (1932), Einstein, Podolsky, and Rosen (1935), Schrödinger (1935,1936) and others. The London-Bauer (1939) theory of measurement and its critique through the story of Wigner's (1961) "friend" are concerned with the possibility already pointed out by von Neumann and Pauli that the observer's consciousness enters in an essential way into the description of quantum measurements. Other denials of the possibility of *realistic* interpretations are formulated in the position that only a *statistical* interpretation of quantum mechanical probabilities is tenable [7,63,141]. In this view quantum mechanics refers only to *ensembles* of measurement outcomes or of physical systems but does *not* lead to statements about properties of *individual* systems. On the other hand, *hidden variable* approaches aimed at restoring *classical* realism in quantum mechanics. These, again, are forced to render quantum mechanics as a statistical theory. Many of such attempts were refuted by a number of no-go-theorems like those by Gleason (1957), Kochen and Specker (1967), or Bell (1966), leaving open up to now only nonlocal, contextual theories such as those of de Broglie (1953), Bohm (1952), or Bohm and Vigier (1954). The "many-worlds interpretation" developed by Everett (1957), DeWitt and Graham (1973) is another way of taking seriously quantum mechanics as a universal theory. We shall be very brief with our subsequent discussions of the early developments (Chapter IV) and refer the interested reader to the monographs of Jammer (1966,1974), and to the collection of papers edited by Wheeler and Zurek (1983).

Reconstructions and generalizations of quantum mechanics (pursued systematically since the 1960s) have aimed at an understanding of the role of Hilbert space in quantum mechanics. One may distinguish three groups of approaches.

(1) The *quantum logic approach* aims at an operational justification of the – generally non-Boolean – structure of the lattice of the propositions of the language of a physical theory [15,107,140,146,168,203]. Measurement theory enters this approach only in an informal way in terms of postulates characterizing propositions as properties of physical systems. In order to establish the formal language of quantum physics, one assumes that elementary propositions are value definite, that is, that there exists an experimental procedure – a measuring process – which shows whether the proposition is true or false. An essential presupposition is that this measuring process will lead to a complete objectification. The importance of the quantum logic approach for the present work lies

in the fact that it supports the attempts at formulating a consistent *realistic* interpretation of quantum mechanics.

(2) The *operational approach* takes as its starting point the *convex structure* of the set of (statistical) *states* representing the preparations of physical experiments. Measurement theoretical aspects are investigated primarily on the object level in terms of the notion of *operation* representing state changes induced by measurements [49,69,102,120,135]. The quantum theory of measurement presented in Chapter III is formulated in the spirit of the operational approach.

(3) The *algebraic approach* emphasizes the algebraic structures of the set of observables and it exploits the formal analogy between classical mechanics and quantum mechanics, aiming, in particular, at convenient "quantization" procedures. One of its advantages is the great formal flexibility which allows for an elegant incorporation of superselection rules and other structural changes generalizing quantum mechanics. Hence this approach offers a mathematical language for a discussion of the measurement problem in more general terms. As a survey and rather exhaustive literature guide the reader may wish to consult the monograph of Primas (1983).

Each of the so-called axiomatic approaches has deepened our understanding of the mathematical and conceptual structures of quantum mechanics. However, none of them led to a thorough justification of the ordinary *Hilbert space quantum mechanics*. In particular, the quantum logical lattice approach is not sufficient for a reconstruction of this theory [116]. Due to this fact, but also due to the success of Hilbert space quantum mechanics, many recent investigations in quantum mechanics have been done directly within the Hilbert space formulation of quantum mechanics. The present work is also written entirely within this framework.

The *recent revival* of interest in foundational issues was encouraged during the 1980s due partly to advances in the formal and conceptual structures of quantum mechanics and also to new experimental possibilities and technological demands. This went hand in hand with new ideas on interpretations and on proposals for solving the objectification problem (see Chapter IV). Fundamental experiments have been performed and these have contributed to bringing the quantum theory of measurement closer to empirical testability. Quantum optical and neutron interferometry experiments on the wave-particle dualism, Einstein–Podolsky–Rosen and delayed choice experiments, macroscopic tunnelling, and mesoscopic quantum effects are some examples. Instead of trying to survey these important developments here, we shall simply refer to the many recent conferences devoted to them such as those in Cologne (1984), Gdańsk (1987), Joensuu (1985, 1987, 1990), Munich (1981), New York (1986), Rome (1989), Tokyo (1983, 1986, 1989) or Vienna (1987).

3. Decision tree

In the *minimal interpretation*, quantum mechanics is regarded as a probabilistic physical theory, consisting of a *language* (propositions about outcomes of measurements), a *probability structure* (a convex set of probability measures representing the possible distributions of measurement outcomes) and *probabilistic laws*. In addition, probabilities are interpreted as limits of relative frequencies of measurement outcomes, that is, in the sense of an *epistemic statistical interpretation*.

It is well-known that the minimal interpretation has not been the only one proposed. Other interpretations were formulated earlier. We shall try to give a fairly systematic list of them along with a sequence of decisions to be made concerning the goals quantum mechanics could be desired to serve. The first decisive question to be answered is the one about the *referent* of quantum mechanics: measurement outcomes (*the epistemic* option) or object systems (*the ontic, or realistic* option)? The ontic answer maintains that quantum mechanics deals with *individual* objects and their properties. It is only here that the measurement problem arises. Following this branch, the second decisive question is the *completeness* of quantum mechanics, that is, the question of whether or not all elements of physical reality can be described by quantum mechanics. The first option leads to, and is motivated by, the consideration of *hidden variable theories* underlying the allegedly *incomplete* theory of quantum mechanics, while as a consequence of the known no-go results one is forced to interpret quantum mechanics as a mere statistical theory about ensembles of objects. In the other option, that of maintaining the *completeness* of quantum mechanics and following a *realistic* interpretation, one is facing the phenomenon of nonobjectivity. Accordingly, quantum mechanical probabilities are *objective* in the sense of propensities, or potentialities, expressing tendencies in the behaviour of individual objects.

Again, in the incompleteness interpretations there is no measurement problem: objectification is not at issue at all, since all properties are considered as real throughout but not as subject to quantum mechanics. Since up to now in the realistic (completeness) interpretation no satisfactory solution of the measurement problems has been found, one is forced into a third decision about the range of validity of quantum mechanics: Is quantum mechanics universally valid or only of limited validity? The current state of knowledge does only allow one to guess the answer to this question. In some sense, the more difficult route is that which maintains the universality. Actually, the many-worlds, witnessing and modal interpretations, as well as those stressing the decisive status of the observer can be read as various attempts to live with the insolubility of the objectification problem. It may finally turn out that the only tenable solution in the spirit of this branch is that of a kind of *unsharp* objectification. Some authors have preferred the other option and concluded that quantum mechanics, originally devised as a theory for microsystems, cannot be extrapolated in a straightforward way to larger systems, such as measuring devices. Thus the reductionistic conviction is given up. In our opinion no conclusive decision between these two options can be made at present. We shall therefore be content to provide a short systematic review of the various approaches to the measurement problem in Chapter IV, guided by the above discussion as summarized in the decision tree of Table 1.

Table 1: Decision Tree: Interpretations of quantum mechanics and approaches to the objectification problem.

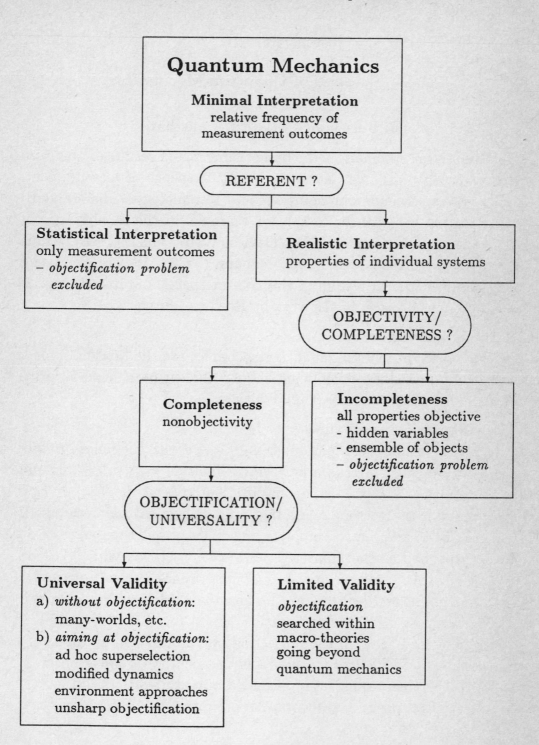

Chapter II

Basic Features of Quantum Mechanics

1. Hilbert space quantum mechanics

This section summarizes the basic elements and results of quantum mechanics which are relevant to the quantum theory of measurement. It also serves to define our notations and terminology. The standard results quoted here can be found, for example, in the monographs of Beltrametti and Cassinelli (1981), Davies (1976), Jauch (1968), Kraus (1983), Ludwig (1983a), and von Neumann (1932). We are also using freely the well-known results of the Hilbert space operator theory, as presented, for instance, in the work by Reed and Simon (1972).

1.1 Basic framework.

The basic concepts of quantum mechanics are the dual notions of states and observables, both being defined in their most general forms in terms of operators acting on a Hilbert space.

a) Mathematical structures.

Let $\mathcal{H}$ be a complex separable Hilbert space with the inner product $\langle\cdot|\cdot\rangle$. An element $\varphi \in \mathcal{H}$ is a unit vector if $\langle\varphi|\varphi\rangle = \|\varphi\|^2 = 1$, and the vectors $\varphi, \psi \in \mathcal{H}$ are orthogonal if $\langle\varphi|\psi\rangle = 0$. A set $\{\varphi_i\} \subset \mathcal{H}$ is orthonormal if the vectors φ_i are mutually orthogonal unit vectors. If $\{\varphi_i\} \subset \mathcal{H}$ is a basis, i.e., a complete orthonormal set, then any $\psi \in \mathcal{H}$ can be expressed as the Fourier series $\psi = \sum\langle\varphi_i|\psi\rangle\varphi_i$ with $\|\psi\|^2 = \sum|\langle\varphi_i|\psi\rangle|^2$. Any unit vector $\varphi \in \mathcal{H}$ determines a one-dimensional projection operator $P[\varphi]$ through the formula $P[\varphi]\psi = \langle\varphi|\psi\rangle\varphi$ for $\psi \in \mathcal{H}$. We also use the bracket notation $|\varphi\rangle\langle\varphi|$ for this projection. If $\{\varphi_i\}$ is a basis of $\mathcal{H}$, then the projection operators $P[\varphi_i]$ are mutually orthogonal and $\sum P[\varphi_i] = I$, where I is the identity operator on $\mathcal{H}$.

Let $\mathcal{L}(\mathcal{H})$ denote the set of bounded operators on $\mathcal{H}$, and $\mathcal{L}(\mathcal{H})^+$ the subset of its positive elements. An operator $A \in \mathcal{L}(\mathcal{H})$ is positive,

$A \geq O$, if $\langle \varphi | A\varphi \rangle \geq 0$ for all vectors $\varphi \in \mathcal{H}$. Then the relation $A \geq B$, defined as $A - B \geq 0$, is an ordering on the subset of self-adjoint bounded operators. Let Ω be a nonempty set and $\mathcal{F}$ a σ-algebra of subsets of Ω so that $(\Omega, \mathcal{F})$ is a measurable space. A normalized positive operator valued (POV) measure $E : \mathcal{F} \to \mathcal{L}(\mathcal{H})^+$ on $(\Omega, \mathcal{F})$ is defined through the properties: $i)$ $E(X) \geq E(\emptyset)$ for all $X \in \mathcal{F}$ (positivity); $ii)$ if (X_i) is a countable collection of disjoint sets in $\mathcal{F}$ then $E(\cup X_i) = \sum E(X_i)$, the series converging in the weak operator topology (σ-additivity); $iii)$ $E(\Omega) = I$ (normalization). For any POV measure $E : \mathcal{F} \to \mathcal{L}(\mathcal{H})^+$ the following two conditions are equivalent: $i)$ $E(X)^2 = E(X)$ for all $X \in \mathcal{F}$; $ii)$ $E(X \cap Y) = E(X)E(Y)$ for all $X, Y \in \mathcal{F}$. Thus a positive operator valued measure is a projection operator valued (PV) measure exactly when it is multiplicative. Further, if the value space $(\Omega, \mathcal{F})$ of E is the real Borel space $(\Re, \mathcal{B}(\Re))$, or a subspace of it, then E determines a unique self-adjoint operator $\int_{\Re} \iota dE$ in $\mathcal{H}$. Here ι denotes the identity function on $\Re$. Conversely, according to the spectral theorem, each self-adjoint operator A in $\mathcal{H}$ defines a unique PV measure $E : \mathcal{B}(\Re) \to \mathcal{L}(\mathcal{H})^+$ such that $A = \int_{\Re} \iota dE$. If E is a PV measure on $(\Re, \mathcal{B}(\Re))$ it shall be denoted as E^A in order to explicate the unique self-adjoint operator A associated with it.

The set of trace class operators on $\mathcal{H}$ will be denoted as $\mathcal{T}(\mathcal{H})$, and $\mathcal{T}(\mathcal{H})_1^+$ consists of the positive trace one operators on $\mathcal{H}$. The trace operation $T \mapsto tr[T]$ is a positive linear functional on $\mathcal{T}(\mathcal{H})$. The one-dimensional projections $P[\varphi]$ are positive operators of trace one. They are the extremal elements of the set $\mathcal{T}(\mathcal{H})_1^+$. Indeed $\mathcal{T}(\mathcal{H})_1^+$ is a convex set (with respect to the linear structure of $\mathcal{T}(\mathcal{H})$), so that an element $T \in \mathcal{T}(\mathcal{H})_1^+$ is extremal if the condition $T = wT_1 + (1-w)T_2$, with $T_1, T_2 \in \mathcal{T}(\mathcal{H})_1^+$, and $0 < w < 1$, always implies that $T = T_1 = T_2$. But $T \in \mathcal{T}(\mathcal{H})_1^+$ is extremal if and only if it is idempotent ($T^2 = T$), which is the case exactly when T is of the form $P[\varphi]$ for some unit vector $\varphi \in \mathcal{H}$. The set of extremal elements of $\mathcal{T}(\mathcal{H})_1^+$ exhaust the whole set $\mathcal{T}(\mathcal{H})_1^+$ in the sense that any $T \in \mathcal{T}(\mathcal{H})_1^+$ can be expressed as a σ-convex combination of some extremal elements $(P[\varphi_i])$: $T = \sum_i w_i P[\varphi_i]$, where (w_i) are suitable weights, that is, $0 \leq w_i \leq 1$, $\sum_i w_i = 1$, and the series converges in the trace norm topology. Such a decomposition can be obtained, in particular, from the spectral decomposition of $T = \sum_i t_i P_i$ (as a positive, self-adjoint compact operator) since the spectral projections P_i (associated with the positive eigenvalues t_i) are finite-dimensional. In that case, T is decomposed into mutually orthogonal one dimensional projections $P[\varphi_i]$, with its eigenvalues representing the weights w_i (each appearing in the series as many times as given by the dimension of the eigenspace).

Let $E : \mathcal{F} \to \mathcal{L}(\mathcal{H})^+$ be a POV measure, and let $T \in \mathcal{T}(\mathcal{H})_1^+$. Then the mapping $E_T : \mathcal{F} \to [0,1], X \mapsto E_T(X) \doteq tr\,[TE(X)]$ is a probability measure. This follows from the defining properties of E and the continuity and linearity of the trace. Finally, the decomposition of states, $T = \sum w_i P[\varphi_i]$, induces the corresponding decomposition of the probability measures $E_T = E_{\sum_i w_i P[\varphi_i]} = \sum_i w_i E_{P[\varphi_i]}$.

b) Physical concepts.

In quantum mechanics a physical system $\mathcal{S}$ is represented by means of a complex separable Hilbert space $\mathcal{H}_{\mathcal{S}}$. The general structure of any experiment – a preparation of a system, followed by a measurement – is reflected in the concepts of states and observables.

The *states* of a system $\mathcal{S}$ are represented by – and identified with – the elements of $\mathcal{T}(\mathcal{H}_{\mathcal{S}})_1^+$. The usual notion of a state as a unit vector of $\mathcal{H}_{\mathcal{S}}$ then refers to the extremal elements of $\mathcal{T}(\mathcal{H}_{\mathcal{S}})_1^+$. We refer to these states $P[\varphi]$, and to the generating unit vectors $\varphi \in \mathcal{H}_{\mathcal{S}}$, as *vector states*. They are often called also *pure states*. In the absence of superselection rules, all vector states are pure states (cf. Section 2.3). Due to the linear structure of $\mathcal{H}_{\mathcal{S}}$, *superpositions* of vector states form new vector states; and any vector state can be represented as a superposition of some other vector states. The convexity of the set of states represents the possibility of preparing new states as *mixtures* of other states.

The notion of an observable provides a representation of the possible events occurring as outcomes of a measurement. In this sense an *observable* is defined as – and identified with – a POV measure $E : \mathcal{F} \to \mathcal{L}(\mathcal{H}_{\mathcal{S}})^+, \; X \mapsto E(X)$ on a measurable space $(\Omega, \mathcal{F})$, the value space of E. Usually, the value space of an observable E is simply (a subspace of) the real Borel space $(\Re, \mathcal{B}(\Re))$, or some of its Cartesian products. The familiar notion of an observable as a self-adjoint opera-

11

tor in $\mathcal{H}_S$ refers to a PV measure on the real line $\Re$. In that case, we shall also use the term *ordinary* observable for both $A = \int_{\Re} \iota d E^A$ as well as the associated spectral measure E^A. In general all PV measures will be called ordinary observables.

A pair (E, T) of an observable E and a state T induces a *probability measure* E_T on the value space $(\Omega, \mathcal{F})$ of E:

$$(1) \qquad E_T : \mathcal{F} \to [0, 1], \ X \mapsto E_T(X) \doteq tr\left[TE(X)\right].$$

The *minimal interpretation* of these probability measures establishes their relation to *measurements* and will be elaborated in Section III.2.4 as the basic ingredient of the notion of measurement.

> MINIMAL INTERPRETATION. *The number $E_T(X)$ is the probability that a measurement of the observable E performed on the system S in the state T leads to a result in the set X.*

This minimal interpretation is contained in any more extensive interpretation of quantum mechanics (cf. Table 1).

In Chapter 1 we mentioned various approaches aiming at a reconstruction of the conceptual structures of quantum mechanics just described. We shall sketch out a combination of arguments coming from the quantum logic and convexity approaches in order to provide some motivation for the notions of states and observables which, in the above generality, may not be too well-known. Starting with the notion of ordinary observables, it is a mathematical consequence of Gleason's theorem (Section 2.1) that states (as linear expectation functionals on the set of bounded self-adjoint operators, or equivalently, as σ-additive probability measures on the lattice of projection operators) are represented by elements of $\mathcal{T}(\mathcal{H}_S)_1^+$ [203]. Next, given $\mathcal{T}(\mathcal{H}_S)_1^+$ as the state space, one may ask for the most general notion of an observable compatible with the probabilistic structure of quantum mechanics [135]. Now the requirement that any state T induces a probability measure on a measurable space $(\Omega, \mathcal{F})$ of measurement outcomes brings about the notion of an observable as a state functional valued measure $X \mapsto E_X$ on $(\Omega, \mathcal{F})$, where $E_X(T)$ represents the probability for an outcome in X in the state T. Assuming that the mapping E_X extends to a linear functional on $\mathcal{T}(\mathcal{H}_S)$ and making use of the fact that the dual space of $\mathcal{T}(\mathcal{H}_S)$ is isomorphic to $\mathcal{L}(\mathcal{H}_S)$, one concludes that to any E_X there corresponds a positive operator $E(X)$ such that $E_X(T) = tr[TE(X)]$. The resulting mapping $X \mapsto E(X)$ is a POV measure. Finally, it turns out (as a slight extension of Gleason's theorem) that this general notion of an observable does not require a restriction, nor does it admit an extension of the notion of state as given above.

In this sense we conclude that the concepts of states and observables as defined here are the most general ones compatible with the probabilistic structure of quantum mechanics. We note that the convexity of the set of states as well as the assumption of the linearity of the mappings E_X can be motivated by statistical characterizations of the assumptions inherent in the notion of state preparation [120,135].

The positive operators in the range of a POV measure E (an observable) represent the events associated with the outcomes which may occur in a measurement of E. They are called *effects*. From the positivity and the normalization of a POV measure it follows that effects are bounded by O and I, that is, $O \leq E(X) \leq I$. Hence their spectrum is a subset of $[0,1]$. An effect $E(X)$ is a projection operator $(E(X) = E(X)^2)$ if and only if its spectrum is the two point set $\{0,1\}$. We let $\mathcal{E}(\mathcal{H}_S) = \{A \in \mathcal{L}(\mathcal{H}_S) : O \leq A \leq I\}$ denote the set of effects. While the projection operators admit an interpretation as *properties* of a physical system, such an interpretation is not possible in the case of general effects. Under certain conditions one may describe effects as *unsharp properties*, constituting *unsharp observables* [37].

The above approach to the notion of an observable leads to a natural generalization of the notion of compatibility: a collection of effects (observables) is *coexistent* if it (the join of their ranges) is contained in the range of some observable. In the case of ordinary observables coexistence is equivalent to commutativity. But in general coexistent pairs of observables need not commute.

We emphasize that the generality inherent in the notion of observable described above is really needed in quantum mechanics. First of all, as just pointed out, this concept allows for the possibility of joint measurements of noncommuting observables, especially complementary observables, in full accordance with the uncertainty relations (see Section III.7.2). This option may also be decisive in an attempt to understand, in terms of quantum mechanics, the almost classical nature of macroscopic observables, like pointers of measuring devices (cf. Section IV.4.4). Further, POV measures provide an appropriate means for dealing with the unsharpness inherent in any real measurement, whatever the sources of the unsharpness may be – quantum mechanical nonobjectivity in microscopic parts of the apparatus, or simply imperfections in the construction of the device. Typical examples of both types of sources, or of their interplay, were exhibited in detailed analyses of Stern-Gerlach experiments for spin observables (for a review, see Ref. 40); on the other hand, the quantum indeterminacy inherent in unsharp position and momentum observables is mandatory for their joint measurability in terms of a quantum mechanical phase space observable (cf. Refs. 49, 102, 176, or 182).

For later use we explicate the structure of the probability measures E_T for discrete observables. An observable E is *discrete* if there is a countable subset Ω_o of the value space Ω such that $E(\Omega \setminus \Omega_o) = O$, the null operator. In that case the probability measures E_T are also discrete and $E_T(X) = \sum_{\omega \in X \cap \Omega_o} E_T(\{\omega\})$. For a vector state $P[\varphi]$ one has

$$E_{P[\varphi]}(\{\omega\}) = tr\,[P[\varphi]E(\{\omega\})] = \langle \varphi | E(\{\omega\})\varphi \rangle.$$

If the value space $(\Omega, \mathcal{F})$ of E is the real Borel space $(\Re, \mathcal{B}(\Re))$, then for an ordinary observable $A = \int_{\Re} \iota dE^A$ one obtains $A = \sum_i a_i E^A(\{a_i\})$, where the numbers a_i are the (disjoint) eigenvalues of A. Let $\{\varphi_{ij}\}$ be a basis of the ith eigenspace of A, so that $A\varphi_{ij} = a_i\varphi_{ij}$ for each j (the index j running from 1 to the dimension of the eigenspace, the degree of degeneracy of a_i). Then the spectral projection $E^A(\{a_i\})$ can be written as $\sum_j P[\varphi_{ij}]$. The bases $\{\varphi_{ij}\}$ of the eigenspaces $E^A(\{a_i\})(\mathcal{H}_S)$, i running over the disjoint eigenvalues of A, form a basis of $\mathcal{H}_S$. Hence we get for the probability $E_T^A(X)$ that a measurement of the observable E^A in the state T leads to a result in the set X:

$$(2) \qquad E_T^A(X) = \sum_{a_i \in X} E_T^A(\{a_i\})$$

$$= \sum_{a_i \in X} tr\left[TE^A(\{a_i\}) \right] = \sum_{ij:a_i \in X} \langle \varphi_{ij} | T \varphi_{ij} \rangle$$

$$= \sum_{a_i \in X} \langle \varphi | E^A(\{a_i\}) \varphi \rangle = \sum_{ij:a_i \in X} |\langle \varphi | \varphi_{ij} \rangle|^2,$$

where the third line refers to the case of $T = P[\varphi]$.

1.2 Tensor product and compound systems.

a) Mathematical structures: tensor product Hilbert spaces.

Let $\mathcal{H}_S$ and $\mathcal{H}_A$ be complex separable Hilbert spaces. The Hilbert space tensor product of $\mathcal{H}_S$ and $\mathcal{H}_A$ is denoted as $\mathcal{H}_S \otimes \mathcal{H}_A$. The linear span of the set $\{\varphi \otimes \Phi : \varphi \in \mathcal{H}_S, \Phi \in \mathcal{H}_A\}$ is dense in $\mathcal{H}_S \otimes \mathcal{H}_A$, and the inner products of the involved Hilbert spaces are related as $\langle \varphi \otimes \Phi | \varphi' \otimes \Phi' \rangle = \langle \varphi | \varphi' \rangle \langle \Phi | \Phi' \rangle$. Moreover, if $\{\varphi_i\}$ and $\{\Phi_k\}$ are bases of $\mathcal{H}_S$ and $\mathcal{H}_A$, then $\{\varphi_i \otimes \Phi_k\}$ is a basis of $\mathcal{H}_S \otimes \mathcal{H}_A$. Any $\Psi \in \mathcal{H}_S \otimes \mathcal{H}_A$ can then be expressed as $\Psi = \sum \langle \varphi_i \otimes \Phi_k | \Psi \rangle \varphi_i \otimes \Phi_k$.

The sets $\mathcal{L}(\mathcal{H}_S \otimes \mathcal{H}_A)$, $\mathcal{T}(\mathcal{H}_S \otimes \mathcal{H}_A)$, etc., and, for example, a POV measure $E : \mathcal{F} \to \mathcal{L}(\mathcal{H}_S \otimes \mathcal{H}_A)^+$ are defined in the usual way. In particular, any $A \in \mathcal{L}(\mathcal{H}_S)$ and $B_A \in \mathcal{L}(\mathcal{H}_A)$ determine a bounded linear operator, their tensor product, $A \otimes B_A$ on $\mathcal{H}_S \otimes \mathcal{H}_A$ via the relation $(A \otimes B_A)(\varphi \otimes \Phi) = A\varphi \otimes B_A\Phi$, $\varphi \in \mathcal{H}_S$, $\Phi \in \mathcal{H}_A$. Also, for any $T \in \mathcal{T}(\mathcal{H}_S)$ and $T_A \in \mathcal{T}(\mathcal{H}_A)$ their tensor product $T \otimes T_A$ is a trace class operator on $\mathcal{H}_S \otimes \mathcal{H}_A$. As is well known, the tensor product operators do not exhaust the respective operator sets. This important fact will be illustrated below in connection with the trace class operators.

In the theory of compound systems the partial trace operation is particularly important. The *partial trace* over the Hilbert space $\mathcal{H}_A$, say, is the positive linear mapping $\Pi_{\mathcal{H}_A} : \mathcal{T}(\mathcal{H}_S \otimes \mathcal{H}_A) \to \mathcal{T}(\mathcal{H}_S)$ defined via the relation

$$(3) \qquad tr\left[\Pi_{\mathcal{H}_A}(W)A\right] = tr[WA \otimes I_A],$$

where $W \in \mathcal{T}(\mathcal{H}_S \otimes \mathcal{H}_A)$, $A \in \mathcal{L}(\mathcal{H}_S)$, and I_A is the identity operator on $\mathcal{H}_A$. If $\{\varphi_i\} \subset \mathcal{H}_S$ and $\{\Phi_k\} \subset \mathcal{H}_A$ are orthonormal bases, then $\Pi_{\mathcal{H}_A}(W)$ can be written as

$$(4) \qquad \Pi_{\mathcal{H}_A}(W) = \sum_{ijk} \langle \varphi_i \otimes \Phi_k | W \varphi_j \otimes \Phi_k \rangle |\varphi_i\rangle\langle\varphi_j|.$$

Here $|\varphi_i\rangle\langle\varphi_j|$ is the bounded linear operator on $\mathcal{H}_S$ given by $|\varphi_i\rangle\langle\varphi_j|(\varphi) = \langle\varphi_j|\varphi\rangle\varphi_i$, $\varphi \in \mathcal{H}_S$. The partial trace over $\mathcal{H}_S$ is defined similarly and denoted as $\Pi_{\mathcal{H}_S} : \mathcal{T}(\mathcal{H}_S \otimes \mathcal{H}_A) \to \mathcal{T}(\mathcal{H}_A)$.

As mentioned above, the operators of the form $T \otimes T_A$ do not exhaust $\mathcal{T}(\mathcal{H}_S \otimes \mathcal{H}_A)$. If $W = T \otimes T_A$, then $T = \Pi_{\mathcal{H}_A}(W)$ and $T_A = \Pi_{\mathcal{H}_S}(W)$, but, in general, $W \neq \Pi_{\mathcal{H}_A}(W) \otimes \Pi_{\mathcal{H}_S}(W)$. In particular, if $W = P[\Psi]$, then $P[\Psi] = \Pi_{\mathcal{H}_A}(P[\Psi]) \otimes \Pi_{\mathcal{H}_S}(P[\Psi])$ if and only if $\Psi = \varphi \otimes \Phi$ for some $\varphi \in \mathcal{H}_S$ and $\Phi \in \mathcal{H}_A$. In that case also $\Pi_{\mathcal{H}_A}(P[\Psi]) = P[\varphi]$ and $\Pi_{\mathcal{H}_S}(P[\Psi]) = P[\Phi]$. This result demonstrates the important fact that the extremal elements of the sets $\mathcal{T}(\mathcal{H}_S)_1^+$ and $\mathcal{T}(\mathcal{H}_A)_1^+$ do not exhaust the set of extremal elements of $\mathcal{T}(\mathcal{H}_S \otimes \mathcal{H}_A)_1^+$.

b) Physical concepts: compound systems.

Consider two physical systems S and A, and assume that they are not identical. Let $\mathcal{H}_S$ and $\mathcal{H}_A$ be the Hilbert spaces on which the descriptions of S and A are based. The description of the compound system $S + A$ is then based on $\mathcal{H}_S \otimes \mathcal{H}_A$ according to the usual ideas of quantum mechanics. The theory of compound systems investigates the connections between the descriptions of the systems S, A, and $S + A$. Here we recall only that any state $W \in \mathcal{T}(\mathcal{H}_S \otimes \mathcal{H}_A)_1^+$ of $S + A$ uniquely determines the states of the subsystems S and A as the partial traces $\Pi_{\mathcal{H}_A}(W)$ and $\Pi_{\mathcal{H}_S}(W)$, respectively. These states are called the *reduced states*. From now on we use the more suggestive notations

$$(5) \qquad \mathcal{R}_S(W) \doteq \Pi_{\mathcal{H}_A}(W) \quad \text{and} \quad \mathcal{R}_A(W) \doteq \Pi_{\mathcal{H}_S}(W)$$

for these states. Thus, for example, $\mathcal{R}_S(W)$ is the reduced state of S if W is the state of $S + A$. The fact that, for example, the reduced state $\mathcal{R}_S(W)$ is uniquely determined by W means, in particular, that the system S can be identified as a subsystem of $S + A$ in the sense that all the probability measures E_T referring to S can be obtained from the probability measures $(E \otimes I_A)_W$ referring to $S + A$; here $E \otimes I_A$ is the POV measure $X \mapsto E(X) \otimes I_A$ and $W \in T(\mathcal{H}_S \otimes \mathcal{H}_A)_1^+$ is such that $T = \mathcal{R}_S(W)$. We re-emphasize the important fact that given a system $S + A$ in a vector state $P[\Psi]$, the subsystems S and A are in vector states $P[\varphi]$ and $P[\Phi]$ if and only if $P[\Psi]$ is a product state and hence equal to $P[\varphi \otimes \Phi] = P[\varphi] \otimes P[\Phi]$. In general the states $\mathcal{R}_S(P[\Psi])$ and $\mathcal{R}_A(P[\Psi])$, $\Psi \in \mathcal{H}_S \otimes \mathcal{H}_A$, $\| \Psi \| = 1$, are no vector states. In such a case one says that the subsystems are correlated or entangled.

1.3 Dynamics.

The time evolution of an isolated system prepared (at time $t = 0$) in a pure state ψ_0 is determined by the *Schrödinger equation*

$$(6) \qquad i\hbar \frac{d}{dt} \psi(t) = H \psi(t) \qquad (\psi(0) = \psi_0),$$

where H is the Hamilton operator of the system. Equivalently, a system prepared (at time $t = 0$) in a state T_0 undergoes the following state change:

$$(7) \qquad T(t) = U^*(t) T_0 U(t) \doteq \mathcal{U}_t(T_0),$$

where $U(t) = exp(\frac{i}{\hbar} Ht)$, $t \in \Re$, is the strongly continuous unitary group induced by the Hamiltonian (via Stone's theorem). The differential form of the above process is given by the *von Neumann–Liouville equation*:

$$(8) \qquad i\hbar \frac{d}{dt} T(t) = HT(t) - T(t)H \doteq \mathcal{L}[T(t)].$$

It has been argued from various sides that part of the problem in understanding quantum measurements is rooted in the fact that the description of the measurement dynamics in terms of a unitary group may be too restrictive (cf. Sections III.6 and IV.4.2). Accordingly, adopting this position, one should look for the most general representation of dynamics by means of families of mappings $\mathcal{V}_t$ on $T(\mathcal{H})$. A fairly general class of state transformations $\mathcal{V}_t$ are the positive, trace preserving linear mappings on $\mathcal{T}(\mathcal{H}_S)$. They are naturally obtained via the concept of *reduced dynamics*: Let $\mathcal{H}_S \otimes \mathcal{H}_A$ be the Hilbert space of a compound system $S + A$, $\mathcal{U}_t$ be the dynamical (unitary) group, and $T \otimes T_A$ some state representing the initial $(t = 0)$ preparation. The time evolution of S, say, is obtained by application of the partial trace at every instant of time, yielding the following family of linear state transformations $\mathcal{V}_t$:

$$(9) \qquad \mathcal{V}_t(T) \doteq \mathcal{R}_S\big(\mathcal{U}_t(T \otimes T_A)\big).$$

These mappings play an important role in the quantum theory of measurement (Chapter III).

In Chapter IV we shall discuss two types of approaches to the measurement problem describing dynamics in terms of families of mappings $\mathcal{V}_t$ applied to the compound system consisting of object system S and apparatus A. In the first type of these approaches the system $S + A$ is extended to a larger system including some *environment* $\mathcal{E}$ such that the unitary dynamics $\mathcal{U}_t$ of $S + A + \mathcal{E}$ gives rise to a nonunitary reduced dynamics $\mathcal{V}_t$ for $S + A$. The second approach is more radical as it suggests that the ordinary unitary dynamics be modified to a more general dynamics $\mathcal{V}_t$: a new linear term added to the von Neumann-Liouville operator $\mathcal{L}$ breaks the *purity* of the unitary mappings, that is, it forces systems initially prepared in pure states to evolve into mixed states.

2. Probabilistic structures of quantum mechanics

Much of the conceptual and interpretational problems of quantum mechanics is rooted in the irreducibility of its probability structure. We shall review those probabilistic aspects of the theory which are particularly important for the measurement problem.

2.1 Gleason's theorem.

A fundamental result in the mathematical foundations of quantum mechanics specifies the class of generalized probability measures on the set $\mathcal{E}(\mathcal{H})$ of effects.

Consider the probability measure $E_T : \mathcal{F} \to [0, 1]$ defined by an observable $E : \mathcal{F} \to \mathcal{L}(\mathcal{H})^+$ and a state $T \in \mathcal{T}(\mathcal{H})_1^+$. According to

Equation (1.1), this measure is a composition of the POV measure E and the state-functional $m_T : \mathcal{L}(\mathcal{H}) \to \mathbf{C}, A \mapsto m_T(A) \doteq tr[TA]$. Since $O \leq E(X) \leq I$ for all $X \in \mathcal{F}$, the numbers $E_T(X) = m_T \circ E(X)$ are, indeed, in the unit interval [0,1]. The measure property of E and the linearity and continuity of m_T imply then that $m_T \circ E$ is a (Kolmogorov) probability measure. In fact the linearity and continuity of the state-functional guarantee that the mapping m_T, when restricted to the set of effects $\mathcal{E}(\mathcal{H})$, is a *generalized probability measure*. That is, it has the following properties. *i)* $m_T(A) \geq m_T(O) = 0$ for all $A \in \mathcal{E}(\mathcal{H})$ (positivity); *ii)* if (A_i) is a countable collection of elements of $\mathcal{E}(\mathcal{H})$ such that $\sum A_i \leq I$, then $m_T(\sum A_i) = \sum m_T(A_i)$ (convergence in the weak operator topology; σ-additivity); *iii)* $m_T(I) = 1$ (normalization). We note that the requirement $\sum A_i \leq I$ in *ii)* is a natural generalization of the condition of pairwise orthogonality of projections. Indeed, if (A_i) is a sequence of projection operators, then $\sum A_i \leq I$ is equivalent to the condition that the A_i are mutually orthogonal, that is, $A_i \leq I - A_j$ for all $i \neq j$.

Let $m : \mathcal{E}(\mathcal{H}) \to [0,1]$ be a generalized probability measure. Then for any observable $E : \mathcal{F} \to \mathcal{L}(\mathcal{H})^+$ the mapping $m \circ E : \mathcal{F} \to [0,1]$ is, again, a probability measure, and the minimal interpretation can be applied to it. The question of whether there are generalized probability measures other than those induced by the states is answered in the negative by Gleason's theorem [82].

THEOREM 2.1.1. *Let $m : \mathcal{E}(\mathcal{H}) \to [0,1]$ be a generalized probability measure. If the vector space dimension of $\mathcal{H}$ is at least 3, then there is exactly one state $T \in \mathcal{T}(\mathcal{H})_1^+$ such that $m(A) = tr[TA]$ for all effects $A \in \mathcal{E}(\mathcal{H})$.*

In its original formulation this theorem referred to probability measures on the lattice of projection operators $\mathcal{P}(\mathcal{H})$; since $\mathcal{P}(\mathcal{H})$ is weakly dense in $\mathcal{E}(\mathcal{H})$, the above considerations show that the probability measures on $\mathcal{E}(\mathcal{H})$ are precisely the extensions of those on $\mathcal{P}(\mathcal{H})$.

Gleason's theorem has several important implications in quantum mechanics. First of all, it specifies the probabilistic content of the state concept. It also ensures that the notions of observables and states as

POV measures and positive trace class operators, respectively, are the most general ones compatible with the probability structure of quantum mechanics. The correspondence between generalized probability measures m_T and states T is one-to-one and onto, and it preserves the natural convex structures of the two sets.

2.2 Irreducibility of probabilities.

Quantum mechanics is an irreducibly probabilistic theory in the following sense: it is impossible to decompose the generalized probability measures into dispersion-free ones, simply because there are no dispersion-free states [15,107]. This important fact is a consequence of Gleason's theorem.

In classical probability theory, the dispersion-free probability measures on a measurable space $(\Omega, \mathcal{F})$ are exactly the extremal elements of the convex set of all probability measures on $(\Omega, \mathcal{F})$. In quantum mechanics the extremal elements of the set of states $\mathcal{T}(\mathcal{H})_1^+$ are precisely the vector states, and none of them corresponds to a dispersion-free generalized probability measure. In fact, to any vector state $P[\varphi]$ there exists an effect A (which may always be chosen to be a projection operator) for which $0 \neq \langle \varphi | A \varphi \rangle \neq 1$. It follows that there are no dispersion-free states in $\mathcal{T}(\mathcal{H})_1^+$.

It is instructive to see under which conditions for a given observable E a probability measure $E_T : \mathcal{F} \to [0,1]$, $T \in \mathcal{T}(\mathcal{H})_1^+$, admits a decomposition into dispersion-free probability measures associated with states. Let $T = \sum w_i P[\varphi_i]$ be a decomposition of T into vector states (φ_i) ($\varphi_i \in \mathcal{H}$, $\| \varphi_i \| = 1$) with weights (w_i), $0 \leq w_i \leq 1$, $\sum w_i = 1$. Then for any $X \in \mathcal{F}$,

$$E_T(X) = \sum w_i E_{P[\varphi_i]}(X) = \sum w_i \langle \varphi_i | E(X) \varphi_i \rangle.$$

Now assume that $\langle \varphi_i | E(X) \varphi_i \rangle \in \{0,1\}$ for all φ_i. This holds true if and only if $E(X)\varphi_i = \varphi_i$ or $E(X)\varphi_i = 0$, which, in turn, is equivalent to the fact that all $P[\varphi_i]$ commute with all $E(X)$. In that case T commutes with all $E(X)$. The converse implication holds true whenever E is an ordinary observable: then $[T, E(X)] = O$ holds exactly if all φ_i are eigenstates of E, so that then $E_T = \sum w_i E_{P[\varphi_i]}$ is a decomposition into dispersion-free probability measures.

2.3 Nonobjectivity of observables.

The irreducibly probabilistic structure of quantum mechanics has an important corollary for the *interpretation* of the basic probabilities $E_T(X)$ of the theory. Even in the case of vector states $P[\varphi]$, the probabilities $\langle\varphi|E(X)\varphi\rangle$ are, as a rule, neither 0 nor 1. In general it is not consistent to assume that in a state $P[\varphi]$ the system S possesses the "property" $E(X)$ or its "complement property" $I - E(X) = E(\Omega\backslash X) = E(X')$, that is, that the value of E in a state $P[\varphi]$ is either in X or in X', respectively. This fact constitutes the *nonobjectivity* of the observable E in the state $P[\varphi]$.

In order to describe the phenomenon of nonobjectivity it is useful to consider first the case of a proper quantum system and then to turn to more general cases. A system S is called a *proper* quantum system if all bounded self–adjoint operators in $\mathcal{L}(\mathcal{H}_S)$ represent (ordinary) observables. In this case the set $\mathcal{O}$ of observables of S is said to be *unrestricted*. In general S may have a *restricted* set of observables, particularly in the presence of a superselection rule. A system S possesses a *superselection rule* if there exists an ordinary observable E which commutes with all other observables $F \in \mathcal{O}$. Such a superselection observable is called a *classical* observable of S. Sometimes it has been suggested that the objects of classical physics are to be characterized as systems the observables of which are all compatible. Accordingly one could consider a *classical* system as a quantum system all observables of which commute with each other.

The possible restriction of the set of observables suggests the introduction of an equivalence relation on the set of states. Two states T_1 and T_2 are *equivalent* with respect to a set $\mathcal{O}$ of observables if these states assign the same probability measures to each observable of $\mathcal{O}$, that is, if $E_{T_1} = E_{T_2}$ for each $E \in \mathcal{O}$. In that case we denote $T_1 \cong_{\mathcal{O}} T_2$, or simply $T_1 \cong T_2$ if there is no ambiguity with regard to the set $\mathcal{O}$. For a given set of observables this relation is, indeed, an equivalence relation on the set $\mathcal{T}(\mathcal{H})_1^+$ of states. No two different states are equivalent with respect to *all* observables unless the set $\mathcal{O}$ of observables is restricted. In the presence of a discrete superselection rule only those vector states represent pure states which are eigenvectors of the superselection observable. All other vector states are equivalent to mixtures of pure states.

Let S be a proper quantum system, A an ordinary observable of S and $X \in \mathcal{B}(\Re)$ any value set of A such that $O \neq E^A(X) \neq I$. We say that $E^A(X)$ corresponds to a *real* property of S in a state T if a measurement of A leads to an outcome in X with a probability equal to one: $E_T^A(X) = 1$. This is the case exactly when $E^A(X)T = T$. Especially for $T = P[\varphi]$ this is equivalent to $E^A(X)\varphi = \varphi$.

Now one can conceive of the following situation. Suppose it is known that either $E^A(X)$ or $E^A(X')$ is a real property of the system S but it is not necessarily known which one is actually the real property. In such a case the properties $E^A(X)$ and $E^A(X')$ are said to be *objective* in the state T of S. But if S had the property $E^A(X)$, or $E^A(X')$, respectively then it should actually be in a state T_X, or $T_{X'}$, respectively in which the property in question *is* real: $E_{T_X}^A(X) = 1$, $E_{T_{X'}}^A(X') = 1$. The given state T should therefore represent a *Gemenge*, that is, a collection $\{(w_X, T_X), (w_{X'}, T_{X'})\}$, where the weights w_X, $w_{X'}$ are the probabilities $w_X = E_T^A(X)$, $w_{X'} = E_T^A(X')$ in the subjective sense of ignorance. Hence, if $E^A(X)$ is objective in the state T, then T admits a decomposition of the form $T = E_T^A(X)T_X + E_T^A(X')T_{X'}$. It follows that $E_T^A(X)T_X = E^A(X)TE^A(X)$ and $E_T^A(X')T_{X'} = E^A(X')TE^A(X')$, and therefore

$$(1) \qquad T = E^A(X)TE^A(X) + E^A(X')TE^A(X')$$

whenever $E^A(X)$ is objective in state T. Condition (1) is, in turn, equivalent to the fact that T commutes with $E^A(X)$.

In accordance with the above considerations we say that an ordinary discrete observable A is *objective* in a state T if all of the properties $E^A(\{a_i\})$ are objective. This is the case only if the state T is a mixture of eigenstates of A. In addition, this mixture must be interpreted as a Gemenge, that is, it must admit an *ignorance interpretation* (see the following Section).

If T is a vector state, $T = P[\varphi]$, then the objectivity of $E^A(X)$ in this state implies, as a consequence of Equation (1), that either $E^A(X)\varphi = \varphi$ or $E^A(X)\varphi = 0$, that is, either $E^A(X)$ or $E^A(X')$ is real in the state $P[\varphi]$ itself. Hence $E^A(X)$ is *nonobjective* in any vector state of a proper quantum system which is not an eigenstate of A. Experimentally, this nonobjectivity becomes manifest in the so-called interference effects. Let a state T_φ be given by the following:

(2) $\qquad T_\varphi \doteq E^A(X)P[\varphi]E^A(X) + E^A(X')P[\varphi]E^A(X').$

Further, let B be another ordinary observable. If the observable A were *objective* in the state $T = P[\varphi]$, then on the basis of equations (1) and (2) one would get

(3) $\qquad tr\left[P[\varphi]E^B(Y)\right] = tr\left[T_\varphi E^B(Y)\right].$

However, without this objectivity assumption one finds the following relationship between the probability measures $E^B_{P[\varphi]}$ and $E^B_{T_\varphi}$:

(4) $\quad tr\left[P[\varphi]E^B(Y)\right] = tr\left[T_\varphi E^B(Y)\right]$
$$+ 2Re(\langle E^A(X)\varphi|E^B(Y)E^A(X')\varphi\rangle).$$

The *interference term* $2Re(\langle E^A(X)\varphi|E^B(Y)E^A(X')\varphi\rangle)$ is zero if φ is an eigenvector of $E^A(X)$. Otherwise it can be made nonzero by a suitable choice of B; in particular, B may not commute with A.

It should be mentioned that this nonobjectivity argument can be sharpened in the following sense. Consider still a proper quantum system $\mathcal{S}$ in the state $T = P[\varphi]$. The premise of the argument is then weakened by assuming merely that one can speak as if either $E^A(X)$ or $E^A(X')$ were a property of the system, without necessarily being objective in the above sense. However, even from this *weak objectivity* assumption about a hypothetical value assignment one can derive a relation which is not compatible with quantum mechanics. Indeed for the probability of another observable B one obtains the following inequality [144].

(5) $\qquad tr\left[P[\varphi]E^B(Y)\right] \leq tr\left[T_\varphi E^B(Y)\right].$

It is obvious that this inequality contradicts the quantum mechanical equation (4) whenever the interference term in (4) is positive. Hence the weak objectivity assumption is not compatible either with quantum mechanics and must thus be refuted (cf. also Section IV.3.3, last paragraph).

Next, if the system $\mathcal{S}$ is not assumed to be a proper quantum system, then the objectivity of $E^A(X)$ in a state $P[\varphi]$ leads to the equivalence $P[\varphi] \cong_{\mathcal{O}} T_\varphi$, with T_φ as given in (2) and $\mathcal{O}$ representing the set of observables of $\mathcal{S}$. If φ is not an eigenvector of $E^A(X)$, then the interference term (4) will not vanish unless the set of observables is restricted. Finally, an observable A is objective in *all* vector states $P[\varphi]$ exactly when A is a classical observable. In particular, all observables of a classical system are objective.

The notion of the objectivity of a property can be generalized for effects as well as for discrete effect valued observables. These generalizations are needed in Section III.5. An effect $E_1 \in \mathcal{E}(\mathcal{H})$ is *real* in a state T if $tr[TE_1] = 1$. The reality of an effect in a state implies that this effect has eigenvalue 1 and that the state in question is a mixture of some 1-eigenstates of the effect. Let $E_2 \doteq I - E_1$ be the "complement" effect of E_1. Effects E_1 and E_2, as well as the simple observable E with $E(\{1\}) = E_1$ and $E(\{2\}) = E_2$, are *objective* in a state T if either E_1 or E_2 is real in this state. The objectivity of E_1 in a state T implies that E_1, as well as E_2, has eigenvalues 0 and 1 and that $T = E_1^1 T E_1^1 + E_2^1 T E_2^1$, where, for example, E_1^1 is the spectral projection of E_1 associated with the eigenvalue 1. Thus

$$(6) \qquad\qquad T = E_1^1 T E_1^1 + E_2^1 T E_2^1$$

whenever E_1 is objective in state T. If (6) holds, then one also has $T = E_1^{1/2} T E_1^{1/2} + E_2^{1/2} T E_2^{1/2}$, which, however, is a weaker relation than (6).

2.4 Nonunique decomposability of mixed states.

Another aspect of the irreducibility of probabilities should be mentioned which is related to the structure of the set of states.

Any mixed state T of $\mathcal{T}(\mathcal{H})_1^+$ admits infinitely many decompositions into vector states $P[\varphi]$ (cf. [15]). One may ask which vector states $P[\varphi]$ can occur as components in some decomposition of T, that is, for which $P[\varphi]$ there exist $w \in (0,1)$ and $T' \in \mathcal{T}(\mathcal{H})_1^+$ such that

$$(7) \qquad\qquad T = wP[\varphi] + (1-w)T'.$$

The answer is as follows: it is precisely the vectors φ in the range of the square root of T which give rise to such a decomposition [93].

This *nonunique decomposability* of mixed states in quantum mechanics constitutes a first indication that such states do not, in general, admit an ignorance interpretation.

2.5 On the ignorance interpretation of mixed states.

The nonunique decomposability of mixed states bears severe implications for the interpretation of such states in quantum mechanics. In fact, generally a mixed state T, with a decomposition $T = \sum w_i P[\varphi_i]$, does *not* admit an *ignorance interpretation* according to which the system $\mathcal{S}$ prepared in state T would actually be in one of the component states $P[\varphi_i]$ with the subjective probabilities w_i. The above result (7) at once makes such an interpretation problematic. However, this question deserves to be studied in greater detail since it is of foremost importance within measurement theory. In order to decide on this issue, let us consider how mixed states can be prepared in quantum mechanics.

Consider a sequence of vector states φ_i, $i = 1, 2, \cdots$, together with a sequence of weights w_i, $0 \le w_i \le 1$, $\sum w_i = 1$. The vectors φ_i may or may not be mutually orthogonal. Assume that a system $\mathcal{S}$ is prepared in such a way that it is known to be in one of the states $P[\varphi_i]$ with subjective probability w_i. This knowledge is represented by assigning to $\mathcal{S}$ the Gemenge $\{(w_i, P[\varphi_i]) : i = 1, 2, \cdots\}$. This Gemenge determines a mixed state $T = \sum w_i P[\varphi_i]$ which represents the probabilistic content of the Gemenge. Due to the knowledge about how the state T was produced one is allowed to apply the ignorance interpretation to the above decomposition. Note, however, that the properties $P[\varphi_i]$ are objective in the above Gemenge only if the vector states φ_i are mutually orthogonal.

Such a Gemenge situation may occur if a preparation instrument, for instance, an accelerator, does not work completely accurately, but it prepares systems in states $\varphi_1, \varphi_2, \cdots$, say, with the *a priori* probabilities

w_1, w_2, $\cdots$, which depend on the construction of the accelerator and which can be determined separately. Thus, if $\mathcal{S}$ is prepared with such an instrument, then $\mathcal{S}$ is to be described by $T = \sum w_i P[\varphi_i]$, which is now the relevant decomposition of T. Whether this type of preparation procedure can also be described within quantum mechanics as a possible physical process will not be analyzed here; see, however, Section III.8.

Another type of situation in which the ignorance interpretation may be applied arises in the presence of a superselection rule. Such a rule may imply that the (mutually orthogonal) vector states φ_1, φ_2, $\cdots$, say, are the only possible pure states of the system. Then the decomposition $T = \sum w_i P[\varphi_i]$ is the only decomposition of T into pure states. This case will be of importance in Section III.5.

A different way of preparing a physical system in a mixed state may occur when the system is part of a compound system. Consider a compound system $\mathcal{S} + \mathcal{A}$ associated with the Hilbert space $\mathcal{H}_\mathcal{S} \otimes \mathcal{H}_\mathcal{A}$. If $\mathcal{S} + \mathcal{A}$ is (prepared) in a vector state Ψ, then the subsystems $\mathcal{S}$ and $\mathcal{A}$ are (prepared) in the reduced states $\mathcal{R}_\mathcal{S}(P[\Psi])$ and $\mathcal{R}_\mathcal{A}(P[\Psi])$ which are vector states $P[\varphi]$ and $P[\Phi]$, say, if and only if $P[\Psi] = P[\varphi] \otimes P[\Phi]$. In a typical measurement situation the state $\mathcal{R}_\mathcal{S}(P[\Psi])$ has a natural decomposition $\mathcal{R}_\mathcal{S}(P[\Psi]) = \sum w_i P[\varphi_i]$ into an orthogonal system of eigenvectors of the measured observable $A = \sum a_i P[\varphi_i]$. Hence, the (necessary) condition of objectivity (1) is satisfied. However, due to the correlations inherent in the state $P[\Psi]$, the observable A cannot be regarded as objective, and the ignorance interpretation cannot be applied to the reduced state of $\mathcal{S}$.

Let $\Psi \in \mathcal{H}_\mathcal{S} \otimes \mathcal{H}_\mathcal{A}$ be a unit vector, which is not of the product form $\Psi = \varphi \otimes \Phi$. Let

$$(8) \qquad \Psi = \sum \sqrt{w_i}\, \varphi_i \otimes \Phi_i$$

be a biorthogonal decomposition (or polar decomposition, or normal form) [107] of Ψ. Now $\mathcal{R}_\mathcal{S}(P[\Psi]) = \sum w_i P[\varphi_i]$. To assume the ignorance interpretation to hold for this decomposition of the state of $\mathcal{S}$ amounts to stating the objectivity of an observable $A \otimes I_\mathcal{A} = \sum a_i P[\varphi_i] \otimes I_\mathcal{A}$ of $\mathcal{S} + \mathcal{A}$ in the state Ψ. Consequently, the state $P[\Psi]$ would have to be equivalent to $T_\Psi = \sum P[\varphi_i] \otimes I_\mathcal{A} P[\Psi] P[\varphi_i] \otimes I_\mathcal{A}$, in accordance with (1), that is, $P[\Psi] \cong T_\Psi$. Clearly this does not hold true unless the set of observables is restricted.

We add that also in the case of a mixed state $\mathcal{R}_\mathcal{S}(P[\Psi])$ which is prepared by separating the system $\mathcal{S}$ from the compound system $\mathcal{S} + \mathcal{A}$ it is not possible to assign a value of the observable A to the system $\mathcal{S}$ hypothetically in the sense of weak objectification. Indeed, if one were to assign a value a_i of A to the system $\mathcal{S}$, one also would assign a value of the observable $A \otimes I_A$ to the compound system $\mathcal{S} + \mathcal{A}$ in the state Ψ. However, then we are confronted with the result mentioned above that even a weak objectification of the observable $A \otimes I_A$ in the state Ψ is not compatible with quantum mechanics since it would lead (in complete analogy to (5)) to the inequality

$$(9) \qquad tr\left[P[\Psi]E^B(Y)\right] \leq tr\left[T_\Psi E^B(Y)\right] .$$

A particular situation arises when the polar decomposition (8) is not unique and the reduced mixed state $\mathcal{R}_\mathcal{S}(P[\Psi])$ is degenerate. Then the assumption of the ignorance interpretation for an observable corresponding to one of the orthogonal decompositions of the object state would again lead to the contradiction mentioned above. However since in the case of a degenerate mixed object state there are infinitely many orthogonal decompositions of this state, one could even assign the values of several observables (corresponding to the different decompositions) to the system. From this somewhat stronger assumption one could derive in addition to inequalities like (9) Bell-type inequalities, which were shown to be violated experimentally, in agreement with quantum mechanics. A well-known example illustrating this situation is given by the singlet state $\Psi = \frac{1}{\sqrt{2}}\{\varphi^+ \otimes \Phi^- - \varphi^- \otimes \Phi^+\}$ of $\mathcal{S} + \mathcal{A}$ consisting of two spin-$\frac{1}{2}$ objects.

Chapter III

The Quantum Theory of Measurement

Survey - The notion of measurement

The purpose of measurements is the determination of properties of the physical system under investigation. In this sense the general conception of measurement is that of an unambiguous comparison: the object system S, *prepared* in a state T, is brought into a suitable contact – a *measurement coupling* – with another, independently prepared system, the *measuring apparatus* from which the *result* related to the measured observable E is *determined* by *reading* the value of the *pointer observable*. It is the goal of the quantum theory of measurement to investigate whether measuring processes, being physical processes, are the subject of quantum mechanics. This question, ultimately, is the question of the universality of quantum mechanics (see Chapter I).

In classical physics all observables are objective in any state, that is, they always assume well-defined though possibly unknown values. Moreover, it is possible in principle to measure them without in any way changing the observed system. Hence the measurement outcome is nothing but the value of the observable *before* as well as *after* the measurement. On the other hand, in the case of quantum mechanical systems for any observable there exist states in which the observable is *not* objective (Section II.2.3). In that case the reading shown by the apparatus cannot refer to an objective value of the observable *before* the measurement. Furthermore, it is not evident that a measurement may be such that its outcome refers to an objective value of the observable *after* the measurement.

In this situation the question arises of how to explain and interpret the occurrence of a particular value of the pointer observable. Answering this question amounts to fixing the notion of measurement. The minimal requirement to be fulfilled by a measurement can be formulated as a *calibration condition*: if the observable to be measured is real, then the measurement should exhibit its value unambiguously and with certainty. We shall show (Section 2.3) that in the quantum theory of measurement for discrete ordinary observables this requirement is equivalent to the *probability reproducibility condition* (Section 1.2). This condition stipulates that the probability measure E_T is "transcribed" into a probability measure for the pointer observable governing the distribution of measurement outcomes in the following sense: if the same E-measurement were repeated sufficiently many times under the same conditions (characterized by T), then $E_T(X)$ would show, in the long run, the *relative frequency* of the occurrence of the measurement results in X.

While the above calibration requirement is generally applicable only to discrete *ordinary* observables (allowing for objective values), the probability reproducibility condition may be and will be taken as the minimal content of the notion of measurement for arbitrary observables. In the way outlined above this minimal concept of measurement is precisely what is needed as the operational basis of the minimal interpretation of quantum mechanical probabilities (cf. Section II.1.1). In Section 2 the measurement theoretical implications of the probability reproducibility condition are worked out, providing, in particular, an analysis of the foundations of the probability interpretation.

To summarize, the probability $E_T(X)$ represents the potential occurrence of a measurement result in X and is *realized* in the course of a measurement in the form of a frequency distribution for the *pointer readings* assumed by the apparatus. The question of how this *pointer objectification* is achieved (in view of the nonobjectivity of the measured observable) constitutes the first part of the so-called *objectification problem* in the quantum theory of measurement (Section 5).

The second part of this problem - *value objectification* - is related to a question touched upon above. A particular pointer reading refers to the object system prior to measurement only if the measured observable was objective before the measurement. Where the observable was nonobjective the question arises as to what happens to the system in the course of the measurement. In the context of the quantum theory of measurement it follows that in general a state change is unavoidable (Section 3.7). Attempts to minimize this irreducible "disturbance" lead to the concept of *ideality* of a measurement. We shall see that this property requires another one: *repeatability*. A *repeatable* measurement will force the system into a state in which the pointer reading X refers to an objective value of the measured observable. This shows that the existence of repeatable measurements is necessary for realistic interpretations of quantum mechanics. For such measurements, pointer objectification entails value objectification via *strong value-correlation*. The probabilistic and information theoretical prerequisites for formulating the objectification problem (Section 5) are investigated in Sections 3 and 4 where the notions of ideal, first kind, and repeatable measurements and their relationships are also analyzed.

To complete and summarize this survey, Section 1 provides an outline of the basic steps of any quantum physical experiment: state preparation and measurement, the latter including registration and reading. We shall also point out the difficulties encountered in an attempt to describe these steps within the quantum theory of measurement. In its conventional formulation measurement theory is restricted to the consideration of measurements which preserve the identity of the object system and the measuring apparatus. Aiming at a description of the measuring apparatus as a quantum mechanical system, measurement theory constitutes part of the theory of compound systems with its own specific questions: how the constitution and preparation of isolated objects is achieved; what types of couplings between the object system and the measuring apparatus may serve as measurements of a given observable of the object system (Section 6); how to determine a result on the apparatus; and how to relate it to the measured observable (Sections 2 through 5); which limitations to the measurability of physical quantities are to be taken into account (Section 7). Finally, the problem of state preparations is revisited in Section 8.

1. General description of measurement

1.1 The problem of isolated systems.

Measurement theory should be able to account for the process of preparation of object systems prior to measurements performed on them. The peculiar nature of quantum mechanical compound systems causes appreciable difficulties in explaining the existence of isolated (that is, dynamically and probabilistically independent) systems, in other words, the existence of *objects*. It may turn out that, strictly speaking, objects can be constituted only in a classical environment since otherwise genuine quantum correlations with the environment would persist.

In fact let $\mathcal{S}+\mathcal{E}$ be an isolated compound system undergoing a unitary dynamical evolution as described in Section II.1.3. The probabilistic independence of $\mathcal{S}$ and $\mathcal{E}$ requires the state of $\mathcal{S}+\mathcal{E}$ to be of product form: $W = T \otimes T_{\mathcal{E}}$. But this form cannot be preserved for all times unless the systems are dynamically independent, that is, $H_{int} = O$, or $\mathcal{U}_t = \mathcal{U}_t^{\mathcal{S}} \otimes \mathcal{U}_t^{\mathcal{E}}$. Therefore, starting with a vector state preparation of the form $\varphi \otimes \psi$, the dynamical evolution will inevitably lead to a *correlated* state $\Psi = U(\varphi \otimes \psi)$, a state which is not of product form. Moreover, this entanglement will persist once the subsystems become dynamically independent. The correlation inherent in such a state manifests itself in the existence of pairs of observables A and $A_{\mathcal{E}}$ which are *strongly correlated* (Section 3). The genuine quantum nature of these correlations is due to the vector nature (purity) of the state and it shows itself in the *nonobjectivity* of A and $A_{\mathcal{E}}$ (Section II.2.3). Such pairs of observables are constructed by making use of the biorthogonal decomposition of the state Ψ, $\Psi = \sum \lambda_i \gamma_i \otimes \eta_i$. Here $\{\gamma_i\}$ and $\{\eta_i\}$ are (or can be extended to) orthonormal bases of the Hilbert spaces of $\mathcal{S}$ and $\mathcal{E}$ and, as such, define pairs of discrete ordinary observables $A = \sum a_i P[\gamma_i]$, $A_{\mathcal{E}} = \sum a_i' P[\eta_i]$ which are, in fact, strongly correlated but not objective in Ψ. It follows that the system $\mathcal{S}$ does not admit an exhaustive description independently of its environment $\mathcal{E}$. On the contrary, any manipulation performed on $\mathcal{E}$ will in general lead to changes on $\mathcal{S}$.

In this situation the only way to break the correlations seems to be the introduction of superselection rules (Section 5). Some authors even conclude that quantum mechanics, rather than being a universal theory, needs to be extended into a more general theory, in particular, because it does not allow for the description of classical systems. If one intends to maintain the universal validity of quantum mechanics, then one seems to be facing the conclusion that both the preparation of objects and their restoration after their interaction with a measuring apparatus are at best only approximately realizable. On the other hand, there are superselection rules, such as the symmetrization rules for compound systems consisting of identical systems, which seem to be valid to a high degree of accuracy. In such cases there is a tendency to incorporate statements concerning the existence of superselection rules into quantum mechanics and to be content with the (possible) universality of the thus extended theory. In our opinion it cannot be decided at present whether the notion of isolated quantum systems really requires the environment to be purely classical. In nature there are only very few interactions "available", so that it may as well be sufficient to have only "a few" superselection rules (referring to classical observables of the environment) in order to avoid entanglement. This possibility can only be decided on the basis of a (not yet existing) theory of the fundamental interactions.

These facts sharpen the difficulties in understanding the objectification of measurement outcomes. Firstly, the formation of correlations between object system and apparatus is necessary for obtaining information in a measurement; but at the same time, objectification requires the restoration of the independence of the systems after the measurement. In the simplest conceivable case, the measurement interaction between an object system S and a measuring apparatus A leads to a state Ψ of the form given above such that the observable to be measured (A) and the pointer observable (A_A) are strongly correlated. The *objectification problem* (Section 5) is then nothing but the question of how the pointer observable can assume a definite value (pointer objectification) which would indicate a value of the measured observable (value objectification). In the present simple model case, the process of objectification, accompanied with the reading of the actual result, would leave the system $S + A$ in one of the product states $\gamma_i \otimes \eta_i$.

It is worth recalling that the pioneers of quantum mechanics were well aware of these problems. There are detailed formal expositions in Pauli (1933), Schrödinger (1935), or von Neumann (1932) and various proposed remedies, like von Neumann's recourse to the observers' consciousness, Heisenberg's "cut", or Bohr's stressing the priority of classical concepts for the description of measurement devices. On the other hand, Einstein, Podolsky and Rosen (1935) tried to base an argument against the completeness of quantum mechanics just on the existence of correlated quantum systems. Some of these, as well as the more recent approaches, will be briefly discussed in Chapter IV.

1.2 Measurement.

Consider an observable E of the object system $\mathcal{S}$. The designing of a measurement of this observable requires specifying a *measuring apparatus* $\mathcal{A}$ with Hilbert space $\mathcal{H}_A$ and a *pointer observable* P_A of $\mathcal{A}$, that is, an observable of $\mathcal{A}$ which is to be correlated with E. It will be useful to allow the value space $(\Omega_A, \mathcal{F}_A)$ of the pointer observable P_A to differ from that of the measured observable. In that case we need, however, a (measurable) function $f : \Omega_A \rightarrow \Omega$ to correlate the value sets of the two observables. We call this function a *pointer function*. Let $T \in \mathcal{T}(\mathcal{H}_S)_1^+$ and $T_A \in \mathcal{T}(\mathcal{H}_A)_1^+$ be the *initial states* of $\mathcal{S}$ and $\mathcal{A}$. The initial state of $\mathcal{S} + \mathcal{A}$ is then $T \otimes T_A$, since we assume that prior to measurement $\mathcal{S}$ and $\mathcal{A}$ are both dynamically and probabilistically independent of each other. A measurement coupling between $\mathcal{S}$ and $\mathcal{A}$ shall be described as a *state transformation* $T \otimes T_A \mapsto V(T \otimes T_A)$ of the compound system $\mathcal{S} + \mathcal{A}$. The final state $V(T \otimes T_A)$ of $\mathcal{S} + \mathcal{A}$ uniquely determines the final states of the subsystems $\mathcal{S}$ and $\mathcal{A}$ as the reduced states $\mathcal{R}_{\mathcal{S}}(V(T \otimes T_A))$ and $\mathcal{R}_A(V(T \otimes T_A))$, respectively. Hence, in particular, the probability measure of the pointer observable P_A in the final state of the measuring apparatus $\mathcal{R}_A(V(T \otimes T_A))$ is completely determined. It is convenient to collect the basic ingredients of the description of measurements into a 5-tuple $\langle \mathcal{H}_A, P_A, T_A, V, f \rangle$.

As pointed out in the Survey, the minimal interpretation of the probability measures E_T requires the *probability reproducibility condition* as the first basic condition which makes the above 5-tuple $\langle \mathcal{H}_A, P_A, T_A, V, f \rangle$ qualify as a measurement of E: the pointer observable P_A, with the function f, and the final state $\mathcal{R}_\mathcal{A}(V(T \otimes T_\mathcal{A}))$ of the measuring apparatus must reproduce the probability measures E_T,

$$
\begin{aligned}
(1) \qquad E_T(X) &= P_{\mathcal{A}, \mathcal{R}_\mathcal{A}(V(T \otimes T_\mathcal{A}))}\big(f^{-1}(X)\big) \\
&= P^f_{\mathcal{A}, \mathcal{R}_\mathcal{A}(V(T \otimes T_\mathcal{A}))}(X)
\end{aligned}
$$

for any value set $X \in \mathcal{F}$, and for all possible initial states $T \in \mathcal{T}(\mathcal{H}_\mathcal{S})^+_1$ of $\mathcal{S}$. The second line in Equation (1) is due to the fact that the pair $(P_\mathcal{A}, f)$ defines another pointer observable $P^f_\mathcal{A}$ (on the value space $(\Omega, \mathcal{F})$) via the relation $P^f_\mathcal{A}(X) = P_\mathcal{A}(f^{-1}(X))$, $X \in \mathcal{F}$. It is a mathematical fact and may not be understood as part of the probability reproducibility condition.

The second basic requirement for $\langle \mathcal{H}_\mathcal{A}, P_\mathcal{A}, T_\mathcal{A}, V, f \rangle$ to qualify as an E-measurement is the *objectification requirement*: the measurement should lead to a definite result. This requirement refers, in the first instance, to the objectivity of the pointer observable in the final state of the measuring apparatus. We shall investigate the objectification requirement of a measurement in Section 5. A 5-tuple $\langle \mathcal{H}_\mathcal{A}, P_\mathcal{A}, T_\mathcal{A}, V, f \rangle$ satisfying the probability reproducibility condition (1) will be called a *premeasurement* of E, in order to emphasize that a *measurement* of E has to fulfil, in addition, the objectification requirement. If the value spaces of the observables E and $P_\mathcal{A}$ are the same, and the function f is the identity function ι on Ω, then we simply write $\langle \mathcal{H}_\mathcal{A}, P_\mathcal{A}, T_\mathcal{A}, V \rangle$ instead of $\langle \mathcal{H}_\mathcal{A}, P_\mathcal{A}, T_\mathcal{A}, V, \iota \rangle$. The premeasurement $\langle \mathcal{H}_\mathcal{A}, P_\mathcal{A}, T_\mathcal{A}, V, f \rangle$ is equivalent to the premeasurement $\langle \mathcal{H}_\mathcal{A}, P^f_\mathcal{A}, T_\mathcal{A}, V \rangle$ in a sense to be explained later. The next three Sections are devoted to an investigation of some properties of premeasurements which are relevant to the problem of objectification.

A concept of measurement which is more general than the one discussed here was proposed by Fine (1969). Instead of using the probability reproducibility condition (1), Fine defines a measurement of an observable E as a 5-tuple $\langle \mathcal{H}_\mathcal{A}, P_\mathcal{A}, T_\mathcal{A}, V, f \rangle$ which has the following property: if T_1 and T_2 are two (initial) states of $\mathcal{S}$, which are inequivalent with respect to E, that is, $E_{T_1} \neq E_{T_2}$, then the (final) reduced states $\mathcal{R}_\mathcal{A}(V(T_1 \otimes T_\mathcal{A}))$ and $\mathcal{R}_\mathcal{A}(V(T_2 \otimes T_\mathcal{A}))$ of $\mathcal{A}$ are inequivalent with respect to $P_{\mathcal{A}}^f$, that is

$$P_{\mathcal{A}, \mathcal{R}_{\mathcal{A}}(V(T_1 \otimes T_{\mathcal{A}}))}(f^{-1}(X)) \neq P_{\mathcal{A}, \mathcal{R}_{\mathcal{A}}(V(T_2 \otimes T_{\mathcal{A}}))}(f^{-1}(X)).$$

The intuitive idea behind this concept of measurement is that a measurement should provide "information concerning the initial state of the object system" [67]. For this purpose the concept will certainly do. But this generality has the consequence that a measurement, in the sense of Ref. 67, does not uniquely single out the measured observable. In particular, there exist observables (namely statistically complete ones, cf. Section 8) the measurement (in Fine's sense) of which amounts to a measurement (in Fine's sense) of *all* observables. In Section 2 it will be shown that the concept of premeasurement based on the probability reproducibility condition (1) is just sufficient to define, in a unique way, the measured observable.

2. Premeasurements

2.1 Basic concepts of measurement theory.

It is a basic result of the quantum theory of measurement that for each observable of a physical system there are premeasurements, and even *normal unitary* premeasurements (Section 2.2). In order to formulate these results and to study some further properties of premeasurements we *assume* that the state transformation $T \otimes T_A \mapsto V(T \otimes T_A)$ induced by the measurement process preserves the convex structure of the set of states. In that case V can be viewed as a trace-preserving positive linear mapping on the state space $\mathcal{T}(\mathcal{H}_S \otimes \mathcal{H}_A)$ of the compound system $\mathcal{S} + \mathcal{A}$. From now on we always assume that the state transformation V in the premeasurement $\langle \mathcal{H}_A, P_A, T_A, V, f \rangle$ is a trace-preserving positive linear mapping, that is a trace-preserving *operation* $V : \mathcal{T}(\mathcal{H}_S \otimes \mathcal{H}_A) \to \mathcal{T}(\mathcal{H}_S \otimes \mathcal{H}_A)$. Such a premeasurement shall be denoted $\mathcal{M}$.

An explicit construction of a premeasurement for discrete ordinary observables has been known since von Neumann's (1932) work. In its present most general form the mentioned existence result is due to

Ozawa (1984). Important intermediate contributions are summarized in the monographs [49] and [120].

Any premeasurement $\mathcal{M}$ of an observable E determines an *instrument*, that is, an operation valued measure $\mathcal{I}_\mathcal{M} : \mathcal{F} \to \mathcal{L}(\mathcal{T}(\mathcal{H}_\mathcal{S}))^+$ through the relation

$$(1) \qquad \mathcal{I}_\mathcal{M}(X)T = \mathcal{R}_\mathcal{S}\left(V(T \otimes T_\mathcal{A}) \cdot I \otimes P_\mathcal{A}(f^{-1}(X))\right)$$

for all $X \in \mathcal{F}, T \in \mathcal{T}(\mathcal{H}_\mathcal{S})$. The instrument $\mathcal{I}_\mathcal{M}$ summarizes all the features of the premeasurement $\mathcal{M}$ that pertain to the object system $\mathcal{S}$. It reproduces the observable E via the equations

$$(2) \qquad E_T(X) = tr\left[\mathcal{I}_\mathcal{M}(X)T\right]$$

for all $X \in \mathcal{F}, T \in \mathcal{T}(\mathcal{H}_\mathcal{S})_1^+$. Further, it gives the non-normalized final states $\mathcal{I}_\mathcal{M}(X)T$ of $\mathcal{S}$, on the condition that the measurement leads to a result in X. In particular, $\mathcal{I}_\mathcal{M}(\Omega)T$ represents the state of $\mathcal{S}$ after the measurement but before reading the result. This interpretation of the instrument $\mathcal{I}_\mathcal{M}$ presupposes, however, that the objectification has been ensured.

In addition to the instrument $\mathcal{I}_\mathcal{M}$, the premeasurement $\mathcal{M}$ of E defines another instrument $\mathcal{I} : \mathcal{F}_\mathcal{A} \to \mathcal{L}(\mathcal{T}(\mathcal{H}_\mathcal{S}))^+$ (on the value space of the pointer observable) via the relation $\mathcal{I}(X_\mathcal{A})T = \mathcal{R}_\mathcal{S}(V(T \otimes T_\mathcal{A}) \cdot I \otimes P_\mathcal{A}(X_\mathcal{A}))$, $X_\mathcal{A} \in \mathcal{F}_\mathcal{A}, T \in \mathcal{T}(\mathcal{H}_\mathcal{S})$ (cf. Equation (1)). This instrument defines, via Equation (2), an observable $\tilde{E}$ with the value space $(\Omega_\mathcal{A}, \mathcal{F}_\mathcal{A})$ such that $\tilde{E} \circ f^{-1} = E$. If such a relation holds between two observables E and $\tilde{E}$, then we say that $\tilde{E}$ is a *refinement* of E, or E is a *coarse-grained version* of $\tilde{E}$. Since $E(\mathcal{F}) \subset \tilde{E}(\mathcal{F}_\mathcal{A})$, the two observables are coexistent (Section II.1.1).

It might seem natural to assume that the measurement mapping $T \otimes T_\mathcal{A} \mapsto V(T \otimes T_\mathcal{A})$ would preserve the extremal points (vector states) of $\mathcal{T}(\mathcal{H}_\mathcal{S} \otimes \mathcal{H}_\mathcal{A})_1^+$, so that V would be a *pure* operation. Since the structure of pure operations is completely known [49], the structure of the corresponding premeasurements could be further exploited. Rather than following this option in general (for this, see Ref. 16), we shall consider a more restricted class of premeasurements – the normal unitary premeasurements – which refer to the notion of complete positivity.

2.2 Unitary and normal unitary premeasurements.

Most treatments of the quantum theory of measurement in the literature are based from the outset on *unitary* premeasurements, for which the measurement coupling $V : T(\mathcal{H}_S \otimes \mathcal{H}_A) \to T(\mathcal{H}_S \otimes \mathcal{H}_A)$ is unitary, that is, $V(W) = U^*WU$, $W \in T(\mathcal{H}_S \otimes \mathcal{H}_A)$, for some unitary operator $U : \mathcal{H}_S \otimes \mathcal{H}_A \to \mathcal{H}_S \otimes \mathcal{H}_A$. In addition, there are usually further specifications made.

A unitary premeasurement $\langle \mathcal{H}_A, P_A, T_A, V, f \rangle$ of an observable E is called a *normal unitary* premeasurement if the following conditions are fulfilled: i) the pointer observable P_A has the same value space and scale as the measured observable E (so that the function f is the identity function); ii) the pointer observable is an ordinary observable, that is, P_A is a PV measure; and iii) the initial state T_A of the measuring apparatus is a vector state $T_A = P[\Phi]$. For a normal unitary premeasurement we use the notation $\mathcal{M}_U \doteq \langle \mathcal{H}_A, P_A, \Phi, U \rangle$, with the understanding that P_A is a PV measure on $(\Omega, \mathcal{F})$ and $f = \iota$. If $\mathcal{M}_U$ is a normal unitary premeasurement of E, then the probability reproducibility condition assumes the simple form

$$(3) \qquad \langle \varphi | E(X) \varphi \rangle = \langle U(\varphi \otimes \Phi) | I \otimes P_A(X) U(\varphi \otimes \Phi) \rangle$$

for all $X \in \mathcal{F}$ and for all $\varphi \in \mathcal{H}_S$, $\| \varphi \| = 1$. Note that the probability reproducibility condition is satisfied for all states T whenever it is satisfied for all vector states φ. This is due to the linearity and continuity of the involved mappings and due to the fact that any state can be expressed as a σ-convex combination of some vector states.

Consider any two premeasurements $\mathcal{M}$ and $\tilde{\mathcal{M}}$ of E, and let $\mathcal{I}_{\mathcal{M}}$ and $\mathcal{I}_{\tilde{\mathcal{M}}}$ be the induced instruments. $\mathcal{M}$ and $\tilde{\mathcal{M}}$ are *equivalent* if the induced instruments are the same: $\mathcal{I}_{\mathcal{M}} = \mathcal{I}_{\tilde{\mathcal{M}}}$. This holds, in particular, for the premeasurements $\langle \mathcal{H}_A, P_A, T_A, V, f \rangle$ and $\langle \mathcal{H}_A, P_A^f, T_A, V \rangle$.

From the outset, the notion of instrument is more general than that given above. An *instrument* is a (normalized) operation valued measure $\mathcal{I} : \mathcal{F} \to \mathcal{L}(T(\mathcal{H}_S))^+$ on a value space $(\Omega, \mathcal{F})$ [49]. Any such instrument determines a unique observable $E : \mathcal{F} \to \mathcal{L}(\mathcal{H}_S)^+$ via the relation $tr[TE(X)] = tr[\mathcal{I}(X)T]$ for all $X \in \mathcal{F}, T \in T(\mathcal{H}_S)$ (cf. Equation (2)). The observable E is the *associate observable* of $\mathcal{I}$. As will be seen below, every observable E on a value space $(\Omega, \mathcal{F})$ is the associate observable of at least one instrument. Such instruments are called *E-compatible*. In fact for each observable E there is a unique family of E-compatible instruments $\mathcal{I}$.

An instrument $\mathcal{I} : \mathcal{F} \to \mathcal{L}\big(\mathcal{T}(\mathcal{H}_S)\big)^+$ is *completely positive* if all the operations $\mathcal{I}(X)$, $X \in \mathcal{F}$, in its range are such. An operation $\phi : \mathcal{T}(\mathcal{H}_S) \to \mathcal{T}(\mathcal{H}_S)$ is completely positive whenever its canonical extension $\phi \otimes \iota$ to $\mathcal{T}(\mathcal{H}_S \otimes \mathbf{C}^n)$ is positive for all $n = 1, 2, \cdots$ (ι denotes the identity operation). An equivalent characterization states that ϕ is completely positive if and only if it can be represented as $\phi(T) = \sum_n A_n T A_n^*$, $T \in \mathcal{T}(\mathcal{H}_S)$, for some countably many bounded linear operators A_n on $\mathcal{H}_S$ [49]. Since completely positive instruments are induced by normal unitary premeasurements [120,157], it is important to recognize that for each observable E there exist E-compatible completely positive instruments. Indeed, let $(X_i)_{i \in I}$ be a countable partition of Ω into disjoint ($\mathcal{F}$-measurable) sets, and let $(T_i)_{i \in I}$ be a collection of states. Then $\mathcal{I}(X)T = \sum_{i \in I} tr\big[TE(X \cap X_i)\big]T_i$ defines an E-compatible instrument which is completely positive.

Consider now a premeasurement $\mathcal{M}$ of an observable E. The instrument $\mathcal{I}_{\mathcal{M}}$ induced by this premeasurement is completely positive whenever the measurement mapping V, or at least its restriction $V|_{\mathcal{T}(\mathcal{H}_S) \otimes [T_{\mathcal{A}}]}$, is completely positive [38]. We summarize the above discussion as follows:

THEOREM 2.2.1. *For any observable E there exists a normal unitary premeasurement $\mathcal{M}_U$ such that*

(4)
$$\langle \varphi | E(X)\varphi \rangle = \langle U(\varphi \otimes \Phi)|I \otimes P_{\mathcal{A}}(X)U(\varphi \otimes \Phi)\rangle$$

for all $X \in \mathcal{F}$, $\varphi \in \mathcal{H}_S$, $\| \varphi \| = 1$. A premeasurement $\tilde{\mathcal{M}}$ of an observable E is equivalent to a normal unitary premeasurement $\mathcal{M}_U$ of E if and only if the induced instrument $\mathcal{I}_{\tilde{\mathcal{M}}}$ is completely positive. An E-compatible instrument is completely positive exactly when it is induced by a premeasurement $\tilde{\mathcal{M}}$ of E for which $\tilde{V}|_{\mathcal{T}(\mathcal{H}_S) \otimes [\tilde{T}_{\mathcal{A}}]}$ is completely positive.

2.3 Discrete ordinary observables.

We review next the basic measurement requirements in the case of discrete ordinary observables. In particular, we shall point out the equivalence of the calibration and probability reproducibility conditions for such observables and then exploit the implications of the latter.

Let $A = \sum a_i E^A(\{a_i\})$ be an ordinary discrete observable, the real numbers a_i being its (distinct) eigenvalues. Let N be the (finite or infinite) number of the eigenvalues of A and $\{\varphi_{ij}\}$ be a complete orthonormal set of eigenvectors of A, $A\varphi_{ij} = a_i\varphi_{ij}$, where for each i the index j runs from 1 to $n(i)$, the dimension of the eigenspace $E^A(\{a_i\})(\mathcal{H}_S)$ (finite or infinite). Thus $\{\varphi_{ij} : i = 1, \cdots, N, j = 1, \cdots, n(i)\}$ is a basis of $\mathcal{H}_S$, and any unit vector $\varphi \in \mathcal{H}_S$ may be expressed as the Fourier series $\varphi = \sum_{ij} c_{ij}\varphi_{ij}$ with the coefficients $c_{ij} = \langle \varphi_{ij}|\varphi \rangle$. For each $X \in \mathcal{B}(\Re)$ we have

$$(5) \qquad E^A_{P[\varphi]}(X) = \langle \varphi|E^A(X)\varphi \rangle = \sum_{ij:a_i \in X} |c_{ij}|^2$$

$$\doteq \sum_{i:a_i \in X} N_i^2 \doteq \sum_{i:a_i \in X} p_\varphi(a_i)$$

where we have adopted the notations $N_i^2 = p_\varphi(a_i) = \langle \varphi|E^A(\{a_i\})\varphi \rangle$ for each $i = 1, \cdots, N$.

Consider a quadruple $\langle \mathcal{H}_A, P_A, \Phi, U \rangle$ consisting of a Hilbert space $\mathcal{H}_A$, a PV measure P_A on $\mathcal{H}_A$, a unit vector $\Phi \in \mathcal{H}_A$, and a unitary mapping $U : \mathcal{H}_S \otimes \mathcal{H}_A \to \mathcal{H}_S \otimes \mathcal{H}_A$. We say that $\langle \mathcal{H}_A, P_A, \Phi, U \rangle$ satisfies the *calibration condition* with respect to a discrete ordinary observable A if the pointer observable P_A shows the reading a_i with certainty whenever this result was certain in the initial state φ of the object system; that is, if

$$(6) \qquad \langle \varphi|E^A(\{a_i\})\varphi \rangle = 1 \quad \text{implies that}$$

$$\langle U(\varphi \otimes \Phi)|I \otimes P_A(\{a_i\})U(\varphi \otimes \Phi) \rangle = 1$$

for all $\varphi \in \mathcal{H}_S$ and for all $i = 1, 2, \cdots$. As an immediate consequence of the linearity and unitarity of U, (6) ensures the validity of the probability reproducibility condition (3).

We may now construct all normal unitary premeasurements of A with a particular choice of the pointer observable $A_{\mathcal{A}} = \int_{\Re} \iota dP_A$. To this end consider a measuring apparatus $\mathcal{A}$ with a Hilbert space $\mathcal{H}_{\mathcal{A}}$ whose dimension is equal to N, the number of distinct eigenvalues of the measured observable. Let $\{\Phi_i : i = 1, \cdots, N\}$ be an orthonormal basis of $\mathcal{H}_{\mathcal{A}}$, and define a pointer observable as the discrete nondegenerate quantity $A_{\mathcal{A}} = \sum_i a_i P[\Phi_i]$ having the same eigenvalues as A. An exhaustive characterization of all normal unitary premeasurements of A with such pointer observables can then be given [16].

THEOREM 2.3.1. *A quadruple* $\langle \mathcal{H}_{\mathcal{A}}, A_{\mathcal{A}}, \Phi, U \rangle$ *with a discrete nondegenerate pointer observable* $A_{\mathcal{A}} = \sum_i^N a_i P[\Phi_i]$ *is a normal unitary premeasurement of the discrete ordinary observable* $A = \sum_{ij} a_i P[\varphi_{ij}]$ *if and only if* U *is a unitary extension of the mapping*

$$(7) \qquad \varphi_{ij} \otimes \Phi \mapsto \psi_{ij} \otimes \Phi_i$$

where $\{\psi_{ij}\}$ *is any set of unit vectors in* $\mathcal{H}_S$ *satisfying the orthogonality conditions* $\langle \psi_{ij} | \psi_{ik} \rangle = \delta_{jk}$ *for all* $j, k = 1, \cdots, n(i)$, *for each* $i = 1, \cdots, N$.

The normal unitary premeasurements of A referred to in the above theorem are very important. They correspond to the unitary measurements of A where the pointer observable is *minimal* in the sense that it is just sufficient to distinguish between the eigenvalues of A. This class of normal unitary premeasurements has important subclasses and it also generates other classes of unitary premeasurements of A. Before describing them we determine the state transformations associated with a normal unitary premeasurement $\mathcal{M}_U^m \doteq \langle \mathcal{H}_{\mathcal{A}}, A_{\mathcal{A}}, \Phi, U \rangle$ of A.

Let $\varphi = \sum c_{ij} \varphi_{ij}$ be the initial vector state of $\mathcal{S}$, so that

$$(8) \qquad U(\varphi \otimes \Phi) = \sum_{ij} c_{ij} \psi_{ij} \otimes \Phi_i$$

is the final state of $\mathcal{S} + \mathcal{A}$. We define

$$(9) \quad \gamma_i = N_i^{-1} \sum_j c_{ij} \psi_{ij}, \ \text{with} \ N_i^2 = \left\| \sum_j c_{ij} \psi_{ij} \right\|^2 = \sum_j |c_{ij}|^2$$

whenever $N_i \neq 0$, and $\gamma_i = 0$ otherwise. Then (8) assumes the form

$$(10) \qquad U(\varphi \otimes \Phi) = \sum_i N_i \gamma_i \otimes \Phi_i,$$

and the final states of $\mathcal{S}$ and $\mathcal{A}$ are:

$$(11) \qquad \mathcal{R}_{\mathcal{S}}(P[U(\varphi \otimes \Phi)]) = \sum_i N_i^2 P[\gamma_i]$$

$$(12) \qquad \mathcal{R}_{\mathcal{A}}(P[U(\varphi \otimes \Phi)]) = \sum_{ik} N_i N_k \langle \gamma_i | \gamma_k \rangle |\Phi_k\rangle\langle\Phi_i|.$$

Furthermore, the instrument $\mathcal{I}_U$ induced by $\mathcal{M}_U^m$ has the form

$$(13) \qquad \mathcal{I}_U(X)P[\varphi] = \sum_{i:a_i \in X} N_i^2 P[\gamma_i]$$

$$= \sum_{i:a_i \in X} \sum_j \sum_l |\psi_{il}\rangle\langle\varphi_{il}|P[\varphi]|\varphi_{ij}\rangle\langle\psi_{ij}|.$$

If U and $\tilde{U}$ are two different unitary extensions of (7), then the resulting $\mathcal{A}$-measurements are equivalent, that is, the instruments $\mathcal{I}_U$ and $\mathcal{I}_{\tilde{U}}$ are the same.

There are three particularly interesting choices of the set $\{\psi_{ij}\}$ in Theorem 2.3.1. First, the set $\{\psi_{ij}\}$ may be an orthonormal system of unit vectors. In that case also the vectors γ_i are orthogonal (for each $\varphi \in \mathcal{H}_{\mathcal{S}}$) and the decomposition (10) is a biorthogonal decomposition (for each $\varphi \in \mathcal{H}_{\mathcal{S}}$). The natural decompositions (11) and (12) of the reduced states of $\mathcal{S}$ and $\mathcal{A}$ are then the spectral ones:

$$(14) \qquad \mathcal{R}_{\mathcal{S}}(P[U(\varphi \otimes \Phi)]) = \sum_i N_i^2 P[\gamma_i]$$

$$(15) \qquad \mathcal{R}_{\mathcal{A}}(P[U(\varphi \otimes \Phi)]) = \sum_i N_i^2 P[\Phi_i].$$

In Section 3.2 we shall see that this choice of the generating vectors ψ_{ij} corresponds exactly to an $\mathcal{A}$-measurement $\mathcal{M}_U^m$ which gives rise to strong correlations between the final component states γ_i and Φ_i of $\mathcal{S}$

and $\mathcal{A}$. Another, more restrictive possibility is that the set $\{\psi_{ij}\}$ is a complete orthonormal set of eigenvectors of A. Then, for each $\varphi \in \mathcal{H}_S$, also $A\gamma_i = a_i\gamma_i$. It turns out (Section 3.4) that this choice of the set $\{\psi_{ij}\}$ is characteristic of measurements $\mathcal{M}_U^m$ which lead to strong correlations between the values of the measured observable and the pointer observable. Finally, the set $\{\psi_{ij}\}$ may be chosen as $\{\varphi_{ij}\}$. A characteristic feature of the related A-measurements is their ideality (Section 3.7). Since this type of (pre)measurements is most often discussed in the literature, we list some of their properties here. Let U_L be a unitary operator on $\mathcal{H}_S \otimes \mathcal{H}_A$ which has the restriction

$$(16) \qquad U_L(\varphi_{ij} \otimes \Phi) = \varphi_{ij} \otimes \Phi_i.$$

The final states of $\mathcal{S}$ and $\mathcal{A}$ are then

$$(17) \qquad \mathcal{R}_S(P[U_L(\varphi \otimes \Phi)]) = \sum_i N_i^2 P[\varphi_{a_i}]$$

$$(18) \qquad \mathcal{R}_A(P[U_L(\varphi \otimes \Phi)]) = \sum_i N_i^2 P[\Phi_i].$$

Here we have

$$\varphi_{a_i} = N_i^{-1} \sum_j c_{ij}\varphi_{ij}$$

if $N_i \neq 0$, and $\varphi_{a_i} = 0$ otherwise. In this case the instrument $\mathcal{I}_L$ is

$$(19) \qquad \mathcal{I}_L(X)P[\varphi] = \sum_{i:a_i \in X} N_i^2 P[\varphi_{a_i}]$$

$$= \sum_{i:a_i \in X} E^A(\{a_i\})P[\varphi]E^A(\{a_i\}).$$

The unitary mapping (16) in the premeasurement $\langle \mathcal{H}_A, A_A, \Phi, U_L \rangle$ of A was already discussed by von Neumann (1932), whereas the instrument (19) induced by that measurement was studied in greater detail by Lüders (1951). In the literature this measurement is called sometimes a Lüders measurement, sometimes a von Neumann–Lüders measurement. Its operational, probabilistic, and information theoretical characterizations will be reviewed in Sections 3 and 4. (For further details, see, for example, Refs. 34, 39 and 49. We choose to call this particular pre-measurement $\langle \mathcal{H}_A, A_A, \Phi, U_L \rangle$ of A its *Lüders (pre-)measurement* and the induced instrument the *Lüders instrument*.

Other classes of unitary premeasurements of (a degenerate) A are obtained from Theorem 2.3.1 by means of the following method: Let B be any refinement of A, so that $A = f(B)$ (or $E^A = E^B \circ f^{-1}$) for some function f. If $\mathcal{M}_U^m$ is a premeasurement of B, then $\langle \mathcal{H}_A, A_A, \Phi, U, f \rangle$ is a premeasurement of A (cf. Section 2.1). In particular, if the premeasurement of B is a Lüders measurement, then the resulting A-measurement will be called a *von Neumann (pre-)measurement*. Accordingly the induced instrument will be referred to as a *von Neumann instrument*. In this view Lüders measurements are but a special class of von Neumann measurements. They turn out to be the ideal ones (cf. Section 3.6). Especially, if B is a maximal (meaning nondegenerate) refinement of A, then the resulting von Neumann instrument of A is

$$(20) \quad \mathcal{I}_{vN}^A(X)P[\varphi] = \sum_{i:a_i \in X} \sum_j P[\varphi_{ij}] P[\varphi] P[\varphi_{ij}]$$

$$= \mathcal{I}_L^B(f^{-1}(X))P[\varphi]$$

$$= \sum_{a_i \in X} \sum_{f(b_{ij})=a_i} E^B(\{b_{ij}\}) P[\varphi] E^B(\{b_{ij}\}).$$

For the sake of clarity, we have indicated here the observable associated with the instrument in question. This measurement was also discussed by von Neumann (1932). It should be noted that, in general, the final state of the object system $\mathcal{S}$ after a von Neumann measurement of A is not the same as the state reached after a Lüders measurement of A.

2.4 Measurement and probability.

According to the minimal interpretation the number $E_T(X)$ is the probability that a measurement of the observable E performed on the system $\mathcal{S}$ in the state T leads to a result in the set X. The probability reproducibility condition requires that the probability measure E_T is transcribed into the probability distribution of the measurement outcomes in a given E-measurement $\mathcal{M}$:

$$E_T(X) = P_{A, \mathcal{R}_A(V(T \otimes T_A))}(f^{-1}(X)).$$

The question arises as to whether it is possible to justify, within the quantum theory of measurement, a statistical interpretation of these probability measures in the following sense: if a measurement of an observable E were repeated, under the same conditions, a sufficient number of times, then the *relative frequency* of outcomes in the set X would approach the number $E_T(X)$. This programme was tackled in the literature through two lines of approach.

One approach (α) is mainly concerned with the formal problems of the relative frequency interpretation [200], leading to a precise formulation and justification of a *measurement statistics* interpretation for quantum mechanics [41]. The advantage of this procedure is the great generality achieved in the sense that measurements of discrete as well as continuous observables are covered.

The second line of thought (β) is related to an idea sketched out in the work of Everett (1957), namely to give a quantum mechanical description of the whole process of collecting measurement statistics. This procedure enables one to prove a quantum mechanical version of the strong law of large numbers, thus providing a justification of the probability interpretation as a statement about a sufficiently large ensemble of systems. In this way a *statistical ensemble* interpretation for quantum mechanics is obtained.

Moreover, this procedure allows one to formalize and analyze all the physical preconditions for the probability interpretation. In fact one task of measurement theory is to determine to what degree of accuracy it is possible to "repeat the same measurement under the same conditions sufficiently many times"; thus it is necessary to investigate the possibilities of preparing "many" identical and independent (object and apparatus) systems in the same state. Some of these questions, especially the problematics of considering identical particles as independent (in view of the symmetrization superselection rule), were studied by Ochs (1980). A limitation on the second approach is that, in its present stage, it is confined to discrete observables and hence to a rather restricted class of unitary measurements.

In the present context it will be sufficient to survey briefly the main results of the two approaches mentioned. To formulate these results in the most general context of an E-measurement $\mathcal{M}$ we shall need to introduce the notion of a reading scale. A *reading scale* is a countable (finite or infinite) partition of the value space of the pointer observable, $\Omega_{\mathcal{A}} = \cup f^{-1}(X_i)$, induced by a countable partition of the value space of the measured observable, $\Omega = \cup X_i$, $X_i \in \mathcal{F}$, $X_i \cap X_j = \emptyset$ for $i \neq j$. Such a reading scale will be denoted $\mathcal{R}$. The notion of a reading scale will also be important in connection with the objectification requirement (Section 5). Indeed it appears that registration and reading of a measurement outcome are always obtained with respect to a given reading scale. A reading scale $\mathcal{R}$ determines a discrete, coarse-grained version of the pointer observable $P_{\mathcal{A}}$,

$$(21) \qquad P_{\mathcal{A}}^{\mathcal{R}} : i \mapsto P_{\mathcal{A},i} \doteq P_{\mathcal{A}}\big(f^{-1}(X_i)\big).$$

The $P_{\mathcal{A}}^{\mathcal{R}}$-value i refers to the pointer reading $f^{-1}(X_i)$ which, in turn, may (or may not) refer to the value X_i of the measured observable E. Clearly, if E is discrete, then there is a natural (finest) reading scale $\mathcal{R}$ such that $P_{\mathcal{A}} = P_{\mathcal{A}}^{\mathcal{R}}$.

α) The first approach towards the measurement statistics interpretation mentioned above refers to the following intended empirical content of the minimal interpretation of the probability measures E_T: if a measurement of E in a state T has been repeated n times, and the result $f^{-1}(X)$ has occurred ν times, then $\lim_{n \to \infty} \nu/n = E_T(X)$. The difficulties encountered in this approach are due to the fact that relative frequencies are not probabilities, and probabilities need not be relative frequencies. A formal justification of this interpretation can however be based on the strong law of large numbers and it amounts to the modal frequency interpretation of probability [200]. When considered within the context of measurement theory, this interpretation allows one to recover all the probabilities $E_T(X)$, $X \in \mathcal{F}$, $T \in \mathcal{T}(\mathcal{H})_1^+$, as the relative frequencies of the measurement outcomes obtained in an E-measurement $\mathcal{M}$ [41]. The *measurement statistics interpretation* of the probabilities $E_T(X)$ with respect to $\mathcal{M}$ constitutes a family of discretized pointer observables $P_{\mathcal{A}}^{\mathcal{R}}$ from which the probabilities

$$(22) \qquad P_{\mathcal{A}, \mathcal{R}_{\mathcal{A}}(V(T \otimes T_{\mathcal{A}}))}^{\mathcal{R}}(\{i\}) = P_{\mathcal{A}, \mathcal{R}_{\mathcal{A}}(V(T \otimes T_{\mathcal{A}}))}\big(f^{-1}(X_i^{(\mathcal{R})})\big)$$
$$= E_T(X_i^{(\mathcal{R})})$$

for any $X_i^{(\mathcal{R})}$ (in $\mathcal{R}$) and for each reading scale $\mathcal{R}$ can be obtained as relative frequencies. That is, for each T and for each $\mathcal{R}$ there is a sequence $\Gamma_{T,\mathcal{R}}$ of $P_A^{\mathcal{R}}$-outcomes such that

$$relf(X_i^{(\mathcal{R})}, \Gamma_{T,\mathcal{R}}) \doteq \lim_{n \to \infty} \frac{1}{n} \sum_{j=1}^{n} c_{X_i^{(\mathcal{R})}}(\Gamma_{T,\mathcal{R}}(j))$$

$$= P_{A,\mathcal{R}_A(V(T \otimes T_A))}^{\mathcal{R}}(\{i\})$$

for each i. Here $c_{X_i^{\mathcal{R}}}$ is the characteristic function of the set $X_i^{\mathcal{R}}$. The fact that for each T the family $\{(\mathcal{R}, \Gamma_{T,\mathcal{R}}) : \mathcal{R} \text{ a reading scale}\}$ forms a good family of special frequency spaces guarantees the consistency of this interpretation [200]. It is to be stressed that for any state T the possible measurement results in an E-measurement $\mathcal{M}$ with respect to a reading scale $\mathcal{R}$ are the elements of a sequence $\Gamma_{T,\mathcal{R}}$ of $P_A^{\mathcal{R}}$-values. They correspond to the values $f^{-1}(X_i)$ of the pointer observable P_A which, in their turn, correspond to the values X_i of the measured observable E. It is another question whether the occurrence of a result i means also that the pointer observable has the value $f^{-1}(X_i)$ after the measurement, or even that E has the value X_i after the measurement. Such qualifications depend on some further properties of the premeasurement $\mathcal{M}$ (Section 5). We summarize this discussion as follows.

THEOREM 2.4.1. *Let $\mathcal{M}$ be a premeasurement of an observable E. For any state T the family $\{(\mathcal{R}, \Gamma_{T,\mathcal{R}}) : \mathcal{R} \text{ a reading scale}\}$ is a good family of frequency spaces associated with the probability space $(\Omega_A, \mathcal{F}_A, P_{A,\mathcal{R}_A(V(T \otimes T_A))})$, so that, in particular, for any $X \in \mathcal{F}$ there is a sequence $\Gamma_{T,\mathcal{R}}$ of $P_A^{\mathcal{R}}$-values such that*

(23) $$relf(X, \Gamma_{T,\mathcal{R}}) = P_{A,\mathcal{R}_A(V(T \otimes T_A))}(f^{-1}(X)) = E_T(X).$$

We may now illustrate the above measurement statistics interpretation of the probability measures E_T in the case of a unitary premeasurement $\mathcal{M}_U^m$ of a discrete ordinary observable $A = \sum a_i E^A(\{a_i\})$ as described in the previous subsection. All the involved probability measures are now discrete, and a natural (finest) reading scale is the one given by the nondegenerate eigenvalues of the pointer observable $A_{\mathcal{A}}$: $\mathcal{R} = \cup\{a_i\}$. (Here we assume that the set $\{a_i : i = 1, \cdots, N\}$ of eigenvalues of the pointer observable $A_{\mathcal{A}}$ is closed.) With respect to this reading scale we have $A_{\mathcal{A}} = A_{\mathcal{A}}^{\mathcal{R}}$, and for each initial state φ of $\mathcal{S}$ there is a sequence Γ_φ of pointer eigenvalues such that

$$(24) \qquad \langle\varphi|E^A(X)\varphi\rangle \;=\; E^{A_{\mathcal{A}}}_{\mathcal{R}_{\mathcal{A}}(P[U(\varphi\otimes\Phi)])}(X) \;=\; relf(X,\Gamma_\varphi)$$

for any $X \in \mathcal{B}(\mathfrak{R})$. We note that in the present case of a discrete ordinary observable A for each initial vector state φ of $\mathcal{S}$ all the probabilities $E^A_{P[\varphi]}(X)$ can indeed be obtained as relative frequencies in *one* sequence Γ_φ of eigenvalues of the pointer observable $A_{\mathcal{A}}$ as indicated in (24).

β) We now turn to the second option to provide a justification of the probability interpretation. In this approach one considers n runs of the same measurement, performed on n identically prepared copies of an object system $\mathcal{S}$, as *one single physical process* to be described by quantum mechanics. Considering this theory as universally valid and complete, one would expect it to be able to predict that in a large system consisting of n equally prepared systems $\mathcal{S}$ the relative frequency of outcomes a_i after a measurement would be almost equal to the quantum mechanical probability. Hence, let $\mathcal{S}^{(n)}$ be an n-body system consisting of n identical copies of $\mathcal{S}$: $\mathcal{S}^{(n)} = \mathcal{S}_1 + \ldots + \mathcal{S}_n$. The associated Hilbert space is the tensor product Hilbert space $\mathcal{H}^{(n)} = \mathcal{H}_1 \otimes \ldots \otimes \mathcal{H}_n$. A premeasurement $\mathcal{M}_U^m$ of A $(= A_1 = \cdots = A_n)$ on $\mathcal{S}$ $(= \mathcal{S}_1 = \cdots = \mathcal{S}_n)$ can now be extended to a premeasurement $\mathcal{M}_U^{m,(n)}$ of the discrete ordinary observable $A^{(n)} = A_1 \otimes \cdots \otimes A_n$ by forming the n-fold tensor products of the constituents of $\mathcal{M}_U^m$. Assuming that the pointer eigenvalues $\{a_i : i = 1, \cdots, N\}$ of $A_{\mathcal{A}}$ of $\mathcal{M}_U^m$ form a closed set, a typical $\mathcal{M}_U^{m,(n)}$-measurement outcome is the n-tuple $(a_{l_1}, \cdots, a_{l_n})$, with $l_k \in \{1, \cdots, N\}$ for each $k = 1, \cdots, n$. Let $l \doteq (l_1, \cdots, l_n)$ and let $\Phi_l^{(n)}$ denote the tensor product $\Phi_{l_1} \otimes \cdots \otimes \Phi_{l_n}$ of the pointer eigenstates Φ_{l_k}. Since the vectors $\Phi_l^{(n)}$ form an orthonormal basis of $\mathcal{H}_{\mathcal{A}}^{(n)}$, one may introduce for each $i = 1, \cdots, N$ a *relative frequency operator* [58,95]

$$(25) \qquad\qquad F_i^{(n)} = \sum_l f^{(n)}(i,l)P[\Phi_l^{(n)}]$$

with the eigenvalues

$$(26) \qquad f^{(n)}(i,l) = \frac{1}{n}\sum_{j=1}^{n}\delta_{l_j,i}$$

The eigenvalue equation

$$F_i^{(n)}\Phi_l^{(n)} = f^{(n)}(i,l)\Phi_l^{(n)}$$

shows that the relative frequency of the pointer value a_i corresponds to a real property in the final component state $\Phi_l^{(n)}$ of the apparatus $\mathcal{A}^{(n)}$, a property which is given by the eigenvalue $f^{(n)}(i,l)$ of $F_i^{(n)}$. This equation can be written equivalently as

$$tr\left[P[\Phi_l^{(n)}]\left(F_i^{(n)} - f^{(n)}(i,l)\right)^2\right] = 0.$$

The probability interpretation, a *statistical ensemble interpretation*, which should be justified here states that if one performed a large number of A-measurements on equally prepared systems, then the relative frequency of the outcomes a_i would approach the probability $p_\varphi(a_i)$. Hence probability is related, again, to the situation after the measuring process, that is, to the (reduced) mixed states $\mathcal{R}_S(P[U(\varphi \otimes \Phi)])$ and $\mathcal{R}_\mathcal{A}(P[U(\varphi \otimes \Phi)])$ of the system S and the apparatus $\mathcal{A}$, respectively. If the initial state of $S^{(n)}$ is $\varphi^{(n)}$, the n-fold tensor product of the vector φ, then the reduced states of $S^{(n)}$ and $\mathcal{A}^{(n)}$ after the $\mathcal{A}^{(n)}$-measurement $M_U^{m,(n)}$ are the n-fold tensor products of the above mixed states, denoted as $T_S^{(n)}$ and $T_\mathcal{A}^{(n)}$, respectively. Though $T_\mathcal{A}^{(n)}$ is not an eigenstate of $F_i^{(n)}$, for large n this state behaves like an eigenstate of the relative frequency operator. Indeed one may prove [57,147] that

$$tr\left[T_\mathcal{A}^{(n)}\left(F_i^{(n)} - p_\varphi(a_i)\right)^2\right] = \frac{1}{n}p_\varphi(a_i)(1 - p_\varphi(a_i)).$$

By this one obtains the following result.

THEOREM 2.4.2. *Let* $T_{\mathcal{A}}^{(n)}$ *be the reduced state of the apparatus* $\mathcal{A}^{(n)}$ *reached after a unitary premeaurement* $\mathcal{M}_U^{m,(n)}$ *performed on the object system* $\mathcal{S}^{(n)}$ *in the state* $\varphi^{(n)}$. *Then for increasing* n *this state is close to being an eigenstate of the relative frequency operators* $F_i^{(n)}$, $i = 1, 2, \cdots, N$, *in the following sense:*

$$(27) \qquad \lim_{n \to \infty} tr \left[T_{\mathcal{A}}^{(n)} \left(F_i^{(n)} - p_\varphi(a_i) \right)^2 \right] = 0.$$

According to this theorem the relative frequency of the pointer value a_i after a premeasurement $\mathcal{M}_U^{m,(n)}$ on an ensemble of n systems $\mathcal{S}$ in states φ approaches the probability $p_\varphi(a_i)$ in the limit of large n. Provided that the reduced state $\mathcal{R}_{\mathcal{A}}(P[U(\varphi \otimes \Phi)])$ can be understood as the description of the Gemenge $\{(p_\varphi(a_i), P[\Phi_i]) : i = 1, \cdots, N\}$, then Theorem 2.4.2 is in fact an important result. It then shows that the number $p_\varphi(a_i)$ can indeed be interpreted as the probability for a_i to be the actual pointer value in the reduced state $\mathcal{R}_{\mathcal{A}}(P[U(\varphi \otimes \Phi)])$ of $\mathcal{A}$. Hence this interpretation of the probability $p_\varphi(a_i)$ would follow from the theory itself, and it would not have to be added as an independent hypothesis. It must, however, be emphasized that this reasoning rests on the applicability of the ignorance interpretation to the reduced apparatus state, which is in conflict with the nonobjectivity argument given in Section II.2.5 (see also Section 5).

A theorem similar to the above one was proved earlier by Hartle (1968) and Graham (1973) and generalized mathematically by Ochs (1977). But instead of referring to the reduced state of the object system or the apparatus after the measurement, these authors prove an analogue of Equation (27) for a vector state preparation prior to the measurement. The relevance of their result to measurement theory therefore seems to be rather limited except when interpretations without objectification are considered (cf. Section IV.3).

If the generating vectors ψ_{ij} in $\mathcal{M}_U^m$ are mutually orthogonal, then the reduced state of the apparatus is a mixture of the eigenvectors of the pointer observable, that is, $\mathcal{R}_{\mathcal{A}}(P[U(\varphi \otimes \Phi)]) = \sum p_\varphi(a_i) P[\Phi_i]$. In this case the above Gemenge interpretation of $\mathcal{R}_{\mathcal{A}}(P[U(\varphi \otimes \Phi)])$ would appear even more suggestive than in the case of a general $\mathcal{M}_U^m$. However, the orthogonality of the vectors ψ_{ij} is still not sufficient for such an interpretation. It may be noted that Theorem 2.4.2 can be generalized to the case of a degenerate (ordinary) pointer observable. Hence this theorem applies to a rather large class of premeasurements, while its interpretation presupposes that a measurement has taken place.

We now come to consider the *realistic* interpretation of the probabilities $p_\varphi(a_i)$ [147]. To this end we return to the case of a premeasurement $\mathcal{M}_U^m$. If the vectors ψ_{ij} are not only orthogonal, but form a complete set of eigenvectors of A, then $\mathcal{M}_U^m$ is a strong value-correlation measurement and therefore repeatable. These properties ensure that the probabilities $p_\varphi(a_i)$ can also be attached to the object system after the measurement, that is, $p_\varphi(a_i) = tr\left[\mathcal{R}_S\left(P[U(\varphi \otimes \Phi)]\right)E^A(\{a_i\})\right]$ for each $i = 1, \cdots, N$ and for any φ. Accordingly the number $p_\varphi(a_i)$ could also be interpreted as the probability for a_i being the actual value of the measured observable A in the final state of the object system, provided that the Gemenge interpretation of $\mathcal{R}_A(P[U(\varphi \otimes \Phi)])$ is justified. Similar remarks apply equally well to the measurement statistics interpretation discussed above.

3. Probabilistic aspects of measurements

In the remaining Sections of Chapter III we shall restrict our attention mainly to normal unitary premeasurements $\mathcal{M}_U$ of a general observable E. As explained in the Survey, a premeasurement qualifies as a *measurement* if it satisfies the *objectification requirement*. This requirement refers to the fact that a measurement leads to a definite result, so that one should be able to "read the actual value" of the pointer observable P_A and to deduce from this the value of the measured observable E. Accordingly the *objectification requirement* can be divided into two parts, *pointer objectification* and *value objectification*, referring to the objectivity of the pointer observable P_A and the measured observable E in the final states $\mathcal{R}_A(V(T \otimes T_A))$ and $\mathcal{R}_S(V(T \otimes T_A))$ of A and S, respectively. Value objectification can be achieved through the pointer objectification via *strong correlations*. Such correlations are not guaranteed by the probability reproducibility condition, but they constitute additional restrictions on the structure of the premeasurements $\mathcal{M}$ of E.

Looked upon in the above way, it is the pointer objectification which is at the core of the measurement problem. In order to find appropriate exact formulations of these questions it will prove useful to study first some correlation properties of premeasurements. Such *probabilistic* properties of measurements can be characterized exclusively in terms of the involved probability measures with no further interpretational assumptions. We shall investigate probabilistic formulations of the measurement theoretical notions mentioned earlier (ideal, first kind, repeatable measurements) and their relationships to the various correlation requirements.

3.1 General formulation of correlations.

A premeasurement $\mathcal{M}$ of an observable E transforms the compound system $\mathcal{S} + \mathcal{A}$ into a correlated state $V(T \otimes T_{\mathcal{A}})$, where T denotes the initial state of $\mathcal{S}$. There are three types of correlations inherent in the final state of $\mathcal{S} + \mathcal{A}$ which seem to be relevant for the measurement process: $i)$ correlations between the component states ; $ii)$ correlations between the observables E and $P_{\mathcal{A}}$; and $iii)$ correlations between the corresponding values of E and $P_{\mathcal{A}}$.

Let $\mu : \mathcal{B}(\Re^2) \to [0,1]$ be a probability measure on the real Borel space $(\Re^2, \mathcal{B}(\Re^2))$, and let μ_1 and μ_2 be its marginal measures so that, for example, $\mu_1(X) = \mu(X \times \Re)$ for each $X \in \mathcal{B}(\Re)$. Assume that the expectations and the variances of μ_i, $i = 1, 2$, are well defined and finite: $\epsilon_i = \int x d\mu_i(x)$, $\vartheta_i = \int (x - \epsilon_i)^2 d\mu_i(x)$. We let σ_i denote the standard deviation of μ_i, that is, $\sigma_i = \sqrt{\vartheta_i}$. The (normalized) *correlation* of μ is then defined as:

$$(1) \qquad \rho(\mu) \doteq \int \frac{(x - \epsilon_1)(y - \epsilon_2)}{\sigma_1 \cdot \sigma_2} d\mu(x, y)$$

(whenever $\sigma_1 \neq 0 \neq \sigma_2$). From Schwarz's inequality one obtains $-1 \leq \rho(\mu) \leq 1$. The marginal measures μ_1 and μ_2 of μ are *uncorrelated* if $\rho(\mu) = 0$, and they are *strongly correlated* if $\rho(\mu) = \pm 1$. The marginals μ_1 and μ_2 are uncorrelated ($\rho(\mu) = 0$) whenever they are independent

($\mu = \mu_1 \times \mu_2$). On the other hand, the case $\rho(\mu) = \pm 1$ of strong correlation occurs if and only if the marginal measures μ_1 and μ_2 are linearly dependent, that is, $\mu_1 = \mu_2^\ell$, where ℓ is a function with the form $\ell(y) = (\sigma_1/\sigma_2)(y \mp \epsilon_2) + \epsilon_1$ and μ_2^ℓ is the induced measure $\mu_2^\ell(X) = \mu_2(\ell^{-1}(X))$ (see, for instance, Ref. 42). In the next three subsections we apply this general notion and characterization to some particular probability measures arising from the premeasurement scheme. In those cases the probability measure μ is always generated as an extension of a mapping of the form $X \times Y \mapsto \langle \Psi | E^A(X) \otimes E^B(Y) \Psi \rangle$ for some self-adjoint operators A (in $\mathcal{H}_S$) and B (in $\mathcal{H}_A$) and for some vector state $\Psi \in \mathcal{H}_S \otimes \mathcal{H}_A$. The marginal measures μ_1 and μ_2 are then directly determined as

$$\mu_1(X) = \mu(X \times \Re) = tr\left[\mathcal{R}_S(P[\Psi])E^A(X)\right]$$
$$\mu_2(Y) = \mu(\Re \times Y) = tr\left[\mathcal{R}_A(P[\Psi])E^B(Y)\right]$$

for $X, Y \in B(\Re)$. Since $\mu = \mu(A, B, \Psi)$ we write $\rho(A, B, \Psi)$ instead of $\rho(\mu(A, B, \Psi))$.

3.2 Strong correlations between component states.

A normal unitary premeasurement $\mathcal{M}_U^m$ of A transforms the compound system $S + A$ as well as the subsystems S and A into final states as given by Equations (8)-(12) of Section 2.3. They are:

$$U(\varphi \otimes \Phi) = \sum N_i \gamma_i \otimes \Phi_i$$
$$\mathcal{R}_S(P[U(\varphi \otimes \Phi)]) = \sum N_i^2 P[\gamma_i]$$
$$\mathcal{R}_A(P[U(\varphi \otimes \Phi)]) = \sum N_i N_k \langle \gamma_i | \gamma_k \rangle |\Phi_k\rangle\langle \Phi_i|$$

With respect to this premeasurement of A the final component states of the object system S and measuring apparatus A are $P[\gamma_i]$ and $P[\Phi_i]$, respectively. The question then arises of the conditions under which these states are strongly correlated. For this case, the relevant probability measure μ is the one determined by the self-adjoint operator $P[\gamma_i] \otimes P[\Phi_i] = P[\gamma_i \otimes \Phi_i]$ and the vector state $U(\varphi \otimes \Phi)$. Since $P[\gamma_i \otimes \Phi_i]$ is a (decomposable) projection operator, the measure μ and its marginals μ_1 and μ_2 are easily determined and one obtains the following expression for the correlation

(2) $\rho(P[\gamma_i], P[\Phi_i], U(\varphi \otimes \Phi)) =$

$$\frac{\langle U(\varphi \otimes \Phi)|P[\gamma_i \otimes \Phi_i]U(\varphi \otimes \Phi)\rangle - tr\,[T'P[\gamma_i]]\,tr\,[T'_{\mathcal{A}}P[\Phi_i]]}{\left\{tr\,[T'P[\gamma_i]]\,(1 - tr\,[T'P[\gamma_i]])\,tr\,[T'_{\mathcal{A}}P[\Phi_i]]\,(1 - tr\,[T'_{\mathcal{A}}P[\Phi_i]])\right\}^{1/2}}$$

Here T' and $T'_{\mathcal{A}}$ denote, for short, the final states of $\mathcal{S}$ and $\mathcal{A}$.

If $\rho(P[\gamma_i], P[\Phi_i], U(\varphi \otimes \Phi)) = 1$ for any $i = 1, \cdots, N$ and for all possible initial vector states φ of $\mathcal{S}$ (for which $0 \neq N_i^2 \neq 1$) we say that $\mathcal{M}_U^m$ is a *strong state-correlation* measurement of A. Such measurements are characterized in the following result [16]:

THEOREM 3.2.1. *Let $\mathcal{M}_U^m$ be a normal unitary premeasurement of A with $U(\varphi_{ij} \otimes \Phi) = \psi_{ij} \otimes \Phi_i$, $i = 1, \cdots, N$, $j = 1, \cdots n(i)$. $\mathcal{M}_U^m$ is a strong state-correlation premeasurement if and only if $\{\psi_{ij} : i = 1, \cdots, N, \ j = 1, \cdots, n(i)\}$ is an orthonormal system.*

The orthogonality of the vectors ψ_{ij} is equivalent to the orthogonality of the vectors γ_i for all initial states φ of $\mathcal{S}$. Hence strong state-correlation implies the orthogonality of the γ_i's, and vice versa.

3.3 Strong correlations between observables.

Let $U(\varphi \otimes \Phi)$ be the final state of $\mathcal{S} + \mathcal{A}$ obtained in a normal unitary premeasurement $\mathcal{M}_U$ of an observable $E : \mathcal{B}(\Re) \to \mathcal{L}(\mathcal{H}_\mathcal{S})^+$, $\varphi \in \mathcal{H}_\mathcal{S}$ being an initial state of $\mathcal{S}$. The probability measure

$$X \times Y \mapsto \langle U(\varphi \otimes \Phi)|E(X) \otimes P_{\mathcal{A}}(Y)U(\varphi \otimes \Phi)\rangle$$

has the marginals

(3) $X \mapsto \langle U(\varphi \otimes \Phi)|E(X) \otimes I_{\mathcal{A}}U(\varphi \otimes \Phi)\rangle = E_{\mathcal{R}_\mathcal{S}(P[U(\varphi \otimes \Phi)])}(X)$

(4) $Y \mapsto \langle U(\varphi \otimes \Phi)|I \otimes P_{\mathcal{A}}(Y)U(\varphi \otimes \Phi)\rangle = P_{\mathcal{A},\mathcal{R}_{\mathcal{A}}(P[U(\varphi \otimes \Phi)])}(Y)$

These measures are simply the probability measures of E and $P_{\mathcal{A}}$ associated with the final states of $\mathcal{S}$ and $\mathcal{A}$. The strong correlation of these marginal measures is thus the strong correlation of the observables E

and $P_\mathcal{A}$ in the final state $U(\varphi \otimes \Phi)$ of $\mathcal{S} + \mathcal{A}$. Due to the probability reproducibility condition,

$$E_{P[\varphi]} = P_{\mathcal{A}, \mathcal{R}_\mathcal{A}}(P[U(\varphi \otimes \Phi)])$$

the strong correlation condition $\rho(E, P_\mathcal{A}, U(\varphi \otimes \Phi)) = 1$ is equivalent to

(5) $$E_{P[\varphi]} = P_{\mathcal{A}, \mathcal{R}_\mathcal{A}}(P[U(\varphi \otimes \Phi)]) = E_{\mathcal{R}_\mathcal{S}(P[U(\varphi \otimes \Phi)])}^{\ell_\varphi}$$

where the function ℓ_φ (which depends on the state φ) is completely determined by the expectations ϵ_i ($i = 1, 2$) and the standard deviations σ_i ($i = 1, 2$) of the two marginal measures. The strong correlation condition (5) presupposes again that $\sigma_1 \neq 0 \neq \sigma_2$.

We say that a premeasurement $\mathcal{M}_U$ of E is a *strong observable-correlation* measurement if

(6) $$\rho(E, P_\mathcal{A}, U(\varphi \otimes \Phi)) = 1$$

for each initial state φ of $\mathcal{S}$ with $\sigma_1 \neq 0 \neq \sigma_2$. This is the case exactly if (5) holds for each $\varphi \in \mathcal{H}_\mathcal{S}$, $\| \varphi \| = 1$, for which $\sigma_1 \neq 0 \neq \sigma_2$.

Consider again a normal unitary premeasurement $\mathcal{M}_U$ of E. Assume that the initial and final probability measures of E, $E_{P[\varphi]}$ and $E_{\mathcal{R}_\mathcal{S}(P[U(\varphi \otimes \Phi)])}$, have always (for each φ) the same first and second moments (so that $\epsilon_1 = \epsilon_2$, $\sigma_1 = \sigma_2$). For such a premeasurement the strong observable-correlation condition (5) assumes the simpler form

(7) $$E_{P[\varphi]} = P_{\mathcal{A}, \mathcal{R}_\mathcal{A}}(P[U(\varphi \otimes \Phi)]) = E_{\mathcal{R}_\mathcal{S}(P[U(\varphi \otimes \Phi)])}$$

This equation turns out to be the condition for a first kind measurement (cf. Section 3.5).

3.4 Strong correlations between values.

We next investigate premeasurements which produce strong correlations between the corresponding values of the measured observable and the pointer observable. To this end, let $\mathcal{M}$ be a premeasurement of an observable E, and let $T \in \mathcal{T}(\mathcal{H}_\mathcal{S})_1^+$ be an initial state of $\mathcal{S}$. Consider a value set $X \in \mathcal{F}$ of E. The corresponding value set of the pointer observable is $f^{-1}(X)$. The premeasurement $\mathcal{M}$ produces strong correlations between the corresponding value sets X and $f^{-1}(X)$ of E and P_A in a state T if

$$\rho\big(E(X), P_A\big(f^{-1}(X)\big), V(T \otimes T_A)\big) = 1,$$

that is, if the marginals

$$Y \mapsto tr\big[\mathcal{R}_S(V(T \otimes T_A))E^{E(X)}(Y)\big]$$
$$Z \mapsto tr\big[\mathcal{R}_A(V(T \otimes T_A))E^{P_A(f^{-1}(X))}(Z)\big]$$

of the probability measure

$$Y \times Z \mapsto tr\big[V(T \otimes T_A)E^{E(X)}(Y) \otimes E^{P_A(f^{-1}(X))}(Z)\big]$$

are strongly correlated.

We say that $\mathcal{M}$ is a *strong value-correlation* measurement if

$$(8) \qquad \rho\big(E(X), P_A(f^{-1}(X)), V(T \otimes T_A)\big) = 1$$

for all $X \in \mathcal{F}$ and $T \in \mathcal{T}(\mathcal{H}_S)_1^+$ (for which the respective standard deviations σ_1 and σ_2 are nonzero). Again, a general characterization for such premeasurements can be given. In particular, if $\mathcal{M}_U$ is a normal unitary premeasurement of E, then (8) implies that all $E(X)$ have two eigenvalues $0 \leq e_0(X) \leq e_1(X) \leq 1$ which need not be 0 nor 1 [42]. Instead of explicating these general conditions and results we shall go on by assuming that E is a discrete ordinary observable.

For a normal unitary premeasurement $\mathcal{M}_U^m$ of $A = \sum a_i E^A(\{a_i\})$ the strong value-correlation condition (8) reads

$$(9) \qquad \rho\big(E^A(\{a_i\}), P[\Phi_i], U(\varphi \otimes \Phi)\big) = 1$$

for each $i = 1, \cdots, N$ and for each $\varphi \in \mathcal{H}_S$, $\|\varphi\| = 1$, for which $0 \neq N_i^2 \neq 1$. Such premeasurements are characterized by the following result [16]:

THEOREM 3.4.1. *Let $\mathcal{M}_U^m$ be a normal unitary premeasurement of A with $U(\varphi_{ij} \otimes \Phi) = \psi_{ij} \otimes \Phi_i$, $i = 1, \cdots, N$, $j = 1, \cdots, n(i)$. $\mathcal{M}_U^m$ is a strong value-correlation measurement if and only if $\{\psi_{ij}\}$ is a complete orthonormal system of eigenvectors of A.*

When compared with Theorem 3.2.1, this theorem shows that strong state-correlation measurements $\mathcal{M}_U^m$ of A are strong value-correlation measurements exactly when the (nonzero) γ_i's are always eigenvectors of A.

3.5 First kind measurements.

The notion of a first kind measurement was discussed by Pauli (1933). Pauli gave two definitions which he considered equivalent. The first definition says that a measurement is of the first kind if it leads to the same result upon repetition. The second definition says that a measurement is of the first kind if the probability of obtaining a particular result is the same both before and after the measurement. In the general context of measurement theory these two definitions are not equivalent. We shall adopt the first definition as the definition of *repeatability* (Section 3.6), and the second as the definition of *first kind measurements*.

A premeasurement $\mathcal{M}$ of an observable E and its associated instrument is said to be of the *first kind* if

$$(10) \qquad \begin{aligned} E_T(X) &= E_{\mathcal{R}_s(V(T \otimes T_A))}(X) \\ &= P_{A, \mathcal{R}_A(V(T \otimes T_A))}(f^{-1}(X)) \end{aligned}$$

for all $X \in \mathcal{F}$ and $T \in \mathcal{T}(\mathcal{H}_S)_1^+$. The second equality in (10) is due to the probability reproducibility condition (1.1) and is not to be considered as part of the definition of first kind measurement. A comparison of (10) with (5) shows that a first kind measurement is always a strong observable-correlation measurement. At the end of Subsection 3.3 we pointed out (sufficient) conditions under which a strong observable-correlation measurement is of the first kind (cf. Equation (7)). In general a premeasurement which produces strong correlations between the measured observable and the pointer observable need not be of the first kind. This can be illustrated by a Lüders measurement $\langle \mathcal{H}_A, A_A, \Phi, U_L \rangle$

of a simple observable $A = a_1 P[\varphi_1] + a_2 P[\varphi_2]$, $a_1 \neq a_2$, in $\mathcal{H}_S = \mathbf{C}^2$. Let U be a unitary mapping on $\mathcal{H}_S$ which permutes the vectors φ_1 and φ_2. Then $\langle \mathcal{H}_A, A_A, \Phi, U \otimes I_A \circ U_L \rangle$ is a strong observable-correlation measurement of A but not a first kind nor a strong value-correlation measurement.

To illustrate the connections between first kind and strong value-correlation measurements, consider a simple POV measure E defined on the two point value set $\{1, 2\}$, with $E(\{i\}) = E_i$. Now $O \leq E_i \leq I$ and $E_2 = I - E_1$. Any E-compatible instrument is generated by two operations ϕ_i $(i = 1, 2)$ with $tr[\phi_i T] = tr[T E_i]$ for all $T \in \mathcal{T}(\mathcal{H}_S)_1^+$. Consider the generalized Lüders operations:

$$(11) \qquad \phi_i T = E_i^{1/2} T E_i^{1/2}, \quad T \in \mathcal{T}(\mathcal{H}_S)_1^+.$$

Since such operations are completely positive (cf. Section 2.2), there is a normal unitary premeasurement $\mathcal{M}_U$ of E which gives rise to such an instrument. One has $\epsilon_1^i = \langle \varphi | E_i \varphi \rangle = tr[\mathcal{R}_A (P[U(\varphi \otimes \Phi)]) E_i] = \epsilon_2^i$ for any $\varphi \in \mathcal{H}_S$, $\| \varphi \| = 1$. Thus this measurement is a first kind measurement. But, as a rule, it is not a strong value-correlation measurement. A short computation shows that it is also a strong value-correlation measurement exactly when $E_i(E_i \varphi) = E_i \varphi$ for all $\varphi \in \mathcal{H}_S$. In that case E_1 as well as E_2 are projection operators.

For a premeasurement $\mathcal{M}_U^m$ of a discrete ordinary observable A the strong value-correlation condition and the first kind property are equivalent. Indeed we have the following result [16]:

THEOREM 3.5.1. *Let $\mathcal{M}_U^m$ be a normal unitary premeasurement of A. $\mathcal{M}_U^m$ is a strong value-correlation measurement if and only if it is of the first kind.*

3.6 Repeatable measurements.

Repeatable measurements are an important class of measurements. In fact, repeatability was already pointed out to be the basis for the value objectification. Intuitively a measurement of an observable is repeatable if its repeated application does not lead to a new result. This idea can be formalized most systematically within the theory of sequential measurements. (See, for example, Refs. 34 or 50.) The premeasurement $\mathcal{M}$ is *repeatable* if its repetition does not lead – from the probabilistic point of view – to a new result, that is, if

$$(12) \qquad tr\left[\mathcal{I}_{\mathcal{M}}(Y)\left(\mathcal{I}_{\mathcal{M}}(X)T\right)\right] \;=\; tr\left[\mathcal{I}_{\mathcal{M}}(Y \cap X)T\right]$$

for all $X, Y \in \mathcal{F}$ and for all $T \in \mathcal{T}(\mathcal{H}_S)_1^+$. We say that an observable E admits a repeatable measurement if there is a premeasurement of E which is repeatable. It is an old issue in the quantum theory of measurement, dating from von Neumann's (1932) work, as to whether observables which admit repeatable measurements are necessarily discrete. On the basis of important contributions by Stinespring (1955), Davies (1976) and others the problem was resolved to a large extent by a result due to Ozawa (1984):

THEOREM 3.6.1. *If an observable E admits a repeatable unitary premeasurement $\mathcal{M}_U$, then E is discrete.*

An observable need not be an ordinary one in order to admit a repeatable unitary measurement. (For an example, see Ref. 34.) We observe next that repeatable measurements are of the first kind. That the converse implication does not hold true in general can be seen by means of a counterexample. The measurement of a simple observable E, introduced by Equation (11), is of the first kind. But, if neither E_1, nor E_2, has eigenvalue 1, this measurement cannot be repeatable. In fact it is never repeatable unless E_1, and thus also E_2, is a projection operator [36]. In the context of ordinary observables the notions of first kind measurement and repeatable measurement coincide [39], as claimed by Pauli (1933).

THEOREM 3.6.2. *Let E be an ordinary observable and $\mathcal{I}$ any of its associated instruments. $\mathcal{I}$ is repeatable if and only if it is of the first kind.*

Obviously a von Neumann measurements of a discrete ordinary observable A is repeatable. But a repeatable measurement of A need not be a von Neumann measurement. Indeed the A-compatible instrument $\mathcal{I}(\{a_i\})T = tr\left[TE^A(\{a_i\})\right]T_i$ of Section 2.2 is repeatable whenever $tr\left[T_iE^A(\{a_i\})\right] = 1$ for each $i = 1, 2, \cdots, N$. Yet it is no von Neumann instrument if some of the eigenvalues a_i of A are degenerate.

3.7 Ideal measurements.

No observable E of a proper quantum mechanical system admits a measurement $\mathcal{M}$ which would leave unchanged *all* the states of the system. This follows immediately from the linearity of the instrument $\mathcal{I}_{\mathcal{M}}$ induced by the measurement $\mathcal{M}$. Hence it is important to investigate to what extent a state change is necessary in a measurement.

The first kind property introduced in Section 3.5 is a probabilistic nondisturbance feature which quantum mechanical measurements may or may not possess. A more stringent nondisturbance property representing a kind of minimal disturbance is that of ideality [15]. A measurement is ideal if it alters the measured system only to the extent which is necessary for the measurement result: all the properties which are real in the initial state of the object system and which are compatible with the measured observable remain real also in the final state of the system.

In the case of ordinary observables A the above intuitive conception of ideality leads to the following *probabilistic* formulation, called here p-ideality. An A-measurement $\mathcal{M}$ and its induced instrument $\mathcal{I}_{\mathcal{M}}$ are called *p-ideal* if for any state $T \in \mathcal{T}(\mathcal{H}_S)_1^+$ and for any property $P \in \mathcal{L}(\mathcal{H}_S)$, $P = P^2 = P^*$, which is compatible with A, the following implication holds true:

(13) $\qquad$ if $\ tr[TP] = 1$, then $\ tr\left[\mathcal{I}_{\mathcal{M}}(\Re)(T)P\right] = 1$.

When applied to the measured observable A itself, the p-ideality of $\mathcal{M}$ implies that for any state T and for all value sets X the following implication holds true:

(14) $\qquad$ if $\ E_T^A(X) = 1$, then $\ E_{\mathcal{I}_{\mathcal{M}}(\Re)T}^A(X) = 1$.

For ordinary observables this condition is equivalent to the repeatability of $\mathcal{M}$. Indeed we have the following result [39]:

THEOREM 3.7.1. *A p-ideal premeasurement of an ordinary observable is repeatable.*

A von Neumann measurement of an ordinary discrete observable is always repeatable, and it satisfies (14). In general such measurements are not p-ideal except when they are Lüders measurements.

As a corollary to the above theorem we note that an ordinary observable A is discrete whenever it admits a p-ideal, unitary (or at least completely positive) measurement (Theorem 3.6.1). In the case of a *discrete* ordinary observable the p-ideality of an A-measurement $\mathcal{M}$ implies the following condition:

$$(15) \qquad \text{if } tr\left[TE^A\left(\{a_i\}\right)\right] = 1, \quad \text{then } \mathcal{I}_{\mathcal{M}}\left(\{a_i\}\right)T = T$$

for all $i = 1, 2, \cdots$ and for all T. Indeed if φ is a vector state for which $\langle\varphi|E^A\left(\{a_i\}\right)\varphi\rangle = 1$, then, by p-ideality, $tr\left[\mathcal{I}_{\mathcal{M}}(\Re)\left(P[\varphi]\right)P[\varphi]\right] = 1$. But in this case the support projection of the state $\mathcal{I}_{\mathcal{M}}(\Re)P[\varphi]$ is contained in $P[\varphi]$, which is possible only if $\mathcal{I}_{\mathcal{M}}(\Re)P[\varphi] = P[\varphi]$. Since $P[\varphi]$ is a vector state with $tr\left[\mathcal{I}_{\mathcal{M}}\left(\{a_i\}\right)P[\varphi]\right] = 1$, we finally have $\mathcal{I}_{\mathcal{M}}\left(\{a_i\}\right)P[\varphi] = P[\varphi]$. By linearity the argument extends to arbitrary states T for which $tr\left[TE^A\left(\{a_i\}\right)\right] = 1$. We shall see below that condition (15) is in fact equivalent to the p-ideality of $\mathcal{I}_{\mathcal{M}}$.

Condition (15) admits an immediate generalization to arbitrary *discrete* observables E. Since this generalization will turn out to be important for the objectification problem we formulate it as the definition of the d-ideality of a measurement, d referring to the discreteness-assumption.

Let E be a discrete observable with the generating (nonzero) effects $E(\{\omega_i\})$, $i = 1, 2, \cdots$. An E-measurement $\mathcal{M}$, or an E-compatible instrument, is *d-ideal* if it does not change the state of $\mathcal{S}$ whenever a particular result is certain from the outset:

$$(16) \qquad \text{if } tr\left[TE(\{\omega_i\})\right] = 1, \quad \text{then } \mathcal{I}_{\mathcal{M}}\left(\{\omega_i\}\right)T = T$$

for all $i = 1, 2, \cdots$ and for any $T \in \mathcal{T}(\mathcal{H}_{\mathcal{S}})_1^+$.

For general discrete observables a d-ideal measurement need not be repeatable nor first kind. Indeed the generalized Lüders instrument $T \mapsto E_i^{1/2}TE_i^{1/2}$ of a discrete observable E (with the generating effects E_1 and E_2) is d-ideal but never repeatable unless the E_i are projection operators. Suppose the E_i have eigenvalue 1 with associated spectral projections E_i^1, and let $U : \mathcal{H}_{\mathcal{S}} \mapsto \mathcal{H}_{\mathcal{S}}$ be a unitary mapping which acts as an identity on the eigenspaces $E_i^1(\mathcal{H}_{\mathcal{S}})$. Then the E-compatible instrument $T \mapsto U E_i^{1/2} T E_i^{1/2} U^{-1}$ is still d-ideal but not first kind, unless U commutes with the E_i.

The question of the structure of ideal, repeatable measurements has been a major issue since von Neumann's (1932) work. These properties are crucial for the realistic interpretation of quantum mechanics insofar as the existence of measurements with these properties ensure the interpretation of possible measurement results as potential properties of the system. In the context of unitary premeasurements, the repeatability of a premeasurement $\mathcal{M}_U$ of an observable E implies its discreteness (Theorem 3.6.1). (We may remind ourselves that in the presently existing proofs the complete positivity of the E-compatible instrument $\mathcal{I}$ must be assumed in order to infer the discreteness of E from the repeatability of $\mathcal{I}$). The d-ideality and repeatability of a measurement $\mathcal{M}$ of a discrete observable E do not imply that E is an ordinary observable. For this an additional property, called nondegeneracy, of the measurement must be postulated. A measurement $\mathcal{M}$ of E is *nondegenerate* if the set of all possible final component states $\{\mathcal{I}_{\mathcal{M}}(X)T : X \in \mathcal{F}, T \in \mathcal{T}(\mathcal{H}_S)_1^+\}$ separates the set of effects; that is, for any $B \in \mathcal{E}(\mathcal{H}_S)$,

(17)   if $tr[B\mathcal{I}_{\mathcal{M}}(X)T] = 0$ for all $X \in \mathcal{F}, T \in \mathcal{T}(\mathcal{H})_1^+$,
        then $B = O$.

The following theorem then holds true [49]:

THEOREM 3.7.2. *A discrete observable E admits a d-ideal, repeatable, nondegenerate measurement if and only if E is an ordinary observable. In that case the premeasurement $\mathcal{M}$ is equivalent to a Lüders measurement of E; that is, the induced instrument $\mathcal{I}_{\mathcal{M}}$ is of the form*

(18) $\mathcal{I}_{\mathcal{M}}(X)T = \displaystyle\sum_{\omega_i \in X} E(\{\omega_i\}) T E(\{\omega_i\})$

for all $T \in \mathcal{T}(\mathcal{H}_S)_1^+$ and for all $X \in \mathcal{F}$.

It can be shown that a d-ideal measurement of a discrete ordinary observable is nondegenerate [34]. Furthermore, a result similar to Theorem 3.7.1 shows that also d-ideality implies repeatability [39]. Hence the results of this Section give rise to the following statement.

COROLLARY 3.7.3. *The d-ideal premeasurements of an ordinary discrete observable are exactly the Lüders measurements.*

A Lüders measurement of a discrete ordinary observable is p-ideal. Thus the concepts of p-ideality and d-ideality are equivalent in the case of ordinary discrete observables. From now on we shall only consider the d-ideality as given by (16), and we refer to it as the *ideality* of a premeasurement.

3.8 Résumé – A classification of premeasurements.

The notion of *premeasurement* covers all conditions to be imposed on a measurement of some observable, *except* the objectification requirement. If a premeasurement is such that the objectification is also achieved, then it is called a *measurement*. In the present Section we summarize the relationships between the various probabilistic characterizations of premeasurement discussed so far. While all of the notions to be listed can be formulated solely in terms of premeasurements, their consideration is actually motivated by the idea of measurements. We therefore drop the prefix "pre" in the sequel.

In the following summary we are using obvious abbreviations for the terms ideality (id), repeatability (rep), first kind property (fk), strong value-correlation (svc), strong observable-correlation (soc) and strong state-correlations (ssc). It is convenient to consider four successive classes of measurements, each containing the subsequent one as a subclass.

A) General observables, general measurements.

Measurements are strong observable-correlation measurements if they are of the first kind; the converse does not hold. Measurements are of the first kind whenever they are repeatable. There exist first kind measurements which are *not* repeatable. Hence we have the implications:

$$(rep) \Longrightarrow (fk) \Longrightarrow (soc).$$

We recall that the existence of a repeatable unitary measurement implies that the associated observable is discrete. We also recall that ideality and repeatability of a (unitary) measurement of some discrete observable do not yet require this observable to be an ordinary one.

B) Ordinary observables.

In addition to the above we note that strong value-correlation measurements are strong observable-correlation measurements.

$$(svc) \Longrightarrow (soc).$$

Moreover, we have the following result:

$$(rep) \Longleftrightarrow (fk).$$

C) Discrete ordinary observables.

$$(id) \Longrightarrow (rep) \Longleftrightarrow (fk).$$

In the case of normal unitary measurements with minimal pointer observable we have, in addition:

$$(rep) \Longleftrightarrow (svc) \Longrightarrow (ssc).$$

As mentioned above, the ideality and repeatability requirements are not sufficient to single out discrete ordinary observables. *Discreteness* follows if repeatability *together with the complete positivity* of some associated instrument is given. For a discrete observable E, besides ideality and repeatability, yet another property, the nondegeneracy, of instruments is needed in order to ensure that E is an *ordinary* observable.

Finally, we recall that, in general, a von Neumann measurement is *not* ideal (though repeatable). In fact ideal measurements of discrete ordinary observables are precisely the Lüders measurements.

D) Nondegenerate, discrete ordinary observable.

Here we note the additional fact:

$$(id) \Longleftrightarrow (rep).$$

We emphasize once more that the the notions of repeatability and ideality are crucial for the realistic interpretation of quantum mechanics. We have seen that quantum mechanics does allow for such measurements. But it is also important to realize that to any observable E belong a large (infinite) class of E-measurements. This fact opens up the possibility of accounting for real (imperfect) measurements.

4. Information theoretical aspects of measurements

The incorporation of the concepts of entropy and information into quantum probability theory runs up against severe difficulties, which are mainly due to the existence of noncommuting observables. There exist only a few contributions (for example, Refs. 65,103,130,150), and no well established quantum information theory has been developed up to now. Nevertheless this problem is of considerable interest in various branches of quantum physics, as demonstrated, for instance, in the monographs of Helstrom (1976), Lindblad (1983) and Thirring (1980). Moreover, some promising results have been obtained in applying information theoretical concepts within the quantum theory of measurement. Information theoretical characterizations of measurements will be the subject of the present section.

Any premeasurement $\mathcal{M}$ of an observable E induces the following transformations

$$(1) \qquad\qquad T \mapsto \mathcal{I}_{\mathcal{M}}(\Omega)T \doteq T_{\Omega},$$

$$(2) \qquad\qquad E_T \mapsto E_{T_{\Omega}}$$

where T is the initial state of the object system $\mathcal{S}$. Moreover, a given reading scale $\mathcal{R}$ – a partition of the value space of E, $\Omega = \cup X_i$, $X_i \in \mathcal{F}$, $X_i \cap X_j = \emptyset$ for $i \neq j$ – defines a natural decomposition of T_{Ω}:

$$(3) \qquad\qquad T_{\Omega} = \sum \mathcal{I}_{\mathcal{M}}(X_i)T \doteq \sum N_i^2 T_{\Omega,i},$$

where $N_i^2 = E_T(X_i)$, and $T_{\Omega,i} = N_i^{-2}\mathcal{I}_{\mathcal{M}}(X_i)T$, if $N_i^2 \neq 0$, and $T_{\Omega,i} = O$ otherwise. Properties of the transformations (1) and (2) reflect the properties of $\mathcal{M}$, and the decomposition (3) is the one for which one might tend to apply the ignorance interpretation. Various probabilistic features of the transformations (1) and (2) were studied in the preceding section. We now come to apply the concepts of entropy and information in order to further characterize these transformations. In particular, we are able to characterize those premeasurements which lead to an optimal separation of the final component states $T_{\Omega,i}$ of the object system $\mathcal{S}$ (Section 4.1). We also specify the conditions under which the deficiency of information (implicit in the probability measure E_T) for predicting a certain measurement outcome can be identified with a potential information gain (associated to E_{T_Ω}) obtained by reading the actual measurement result (Section 4.2). Some special information theoretical aspects of von Neumann and Lüders measurements will be pointed out as illustrations of the general relations obtained. Finally, an information theoretical characterization of the commutativity of discrete ordinary observables will be discussed in Section 4.3.

4.1 The concept of entropy.

Following von Neumann (1932), the *entropy of a state T* is defined as the (non-negative) number $S(T) \doteq -tr[Tln(T)]$. Using the spectral decomposition $T = \sum t_i E^T(\{t_i\})$ of T, one obtains the expression

(4) $$S(T) = -\sum_i t_i ln(t_i) \ tr\left[E^T(\{t_i\})\right].$$

Since all the spectral projections $E^T(\{t_i\})$ (with $t_i \neq 0$) are finite dimensional, we may also write

(5) $$S(T) = -\sum_i t_i ln(t_i)$$

whenever T is nondegenerate, or if we allow each term $t_i ln(t_i)$ to appear in the series as many times as indicated by the degeneracy of the eigenvalue t_i. If $Tln(T)$ is a trace class operator, then $S(T)$ is finite. In particular, if the dimension of the range of T is finite, say n, then $0 \leq S(T) \leq ln(n)$. The case $S(T) = 0$ occurs exactly when T is a vector state, whereas $S(T) = ln(n)$ holds if and only if T is totally degenerate (maximally mixed) [66].

Consider any decomposition of a state T into some other states T_i, $T = \sum w_i T_i$, with $0 \leq w_i \leq 1$, $\sum w_i = 1$, as in Equation (3). The concavity and the subadditivity of the entropy functional give rise to the following inequalities.

(6) $$\sum w_i S(T_i) \leq S\left(\sum w_i T_i\right) = S(T) \leq \sum w_i S(T_i) + S(\{w_i\})$$

Here we use the notation $S(\{w_i\}) \doteq -\sum w_i ln(w_i)$. The lower bound $S(T) = \sum w_i S(T_i)$ is obtained if and only if all the component states T_i

are the same, that is $T_i = T_j$ for all i and j, whereas the upper bound $S(T) = \sum w_i S(T_i) + S(\{w_i\})$ is obtained exactly when the component states T_i are mutually disjoint, that is $T_i T_j = O$ for all $i \neq j$ [130]. Thus, in particular, if $T = \sum w_i T_i$ is an orthogonal decomposition of T into vector states $T_i = P[\gamma_i]$, then $S(T) = S(\{w_i\})$, as in Equation (5).

In an intuitive sense the entropy of a state T is a measure of the degree of "mixedness" of T. In the case of an orthogonal decomposition of T into vector states, $T = \sum w_i P[\gamma_i]$, one might ask whether the number $S(T) = S(\{w_i\})$ could be interpreted epistemically as the (average) deficiency of information on the actual vector state γ_i of the system if the system is known to be in the mixed state T. However, such an interpretation seems to require the Gemenge interpretation of T with respect to its decomposition $\sum w_i P[\gamma_i]$. The nonobjectivity arguments of Section II.2.5 show that such an interpretation cannot be justified in general. However, we tentatively assume that in a measurement context the ignorance interpretation can indeed be applied to the final state T_Ω of the object system with respect to its natural decomposition (3). This assumption is necessary for the physical interpretation of the subsequent considerations. The validity of the corresponding formal results does not, of course, depend on interpretational issues.

Consider a measurement $\mathcal{M}$ of an observable E. The state change $T \mapsto T_\Omega$ induced by $\mathcal{M}$ (Equation (1)) is accompanied with a change of entropy, $S(T) \mapsto S(T_\Omega)$. Now pick a reading scale $\mathcal{R}$. Then (3) is the decomposition of T_Ω with respect to this reading scale, and the inequalities (6) give bounds to the entropy $S(T_\Omega)$ for this decomposition. We call a measurement $\mathcal{M}$ a *maximal state entropy measurement* with respect to a reading scale $\mathcal{R}$ if

$$(7) \qquad S(T_\Omega) = S\left(\sum_i N_i^2 T_{\Omega,i}\right) = \sum_i N_i^2 S(T_{\Omega,i}) + S(\{N_i^2\})$$

for any initial state T of $\mathcal{S}$. On account of (6) $\mathcal{M}$ is a maximal state entropy measurement with respect to $\mathcal{R}$ if and only if the component state $T_{\Omega,i}$ are mutually orthogonal for any T. In such a case the measurement leads to an optimal separation of the final component states.

Let $\mathcal{M}$ be a repeatable measurement of E (Section 3.6). Then for any state T and for any value set X (with $E_T(X) \neq 0$) the state

$E_T(X)^{-1}\mathcal{I}_\mathcal{M}(X)T$ is an eigenstate of $E(X)$ associated with its eigenvalue 1. For any disjoint value sets X and Y, the effects $E(X)$ and $E(Y)$ are also disjoint, that is, $E(X) + E(Y) \leq I$, so that the associated 1-eigenstates are orthogonal. We thus obtain the following result.

THEOREM 4.1.1. *A repeatable premeasurement* $\mathcal{M}$ *of an observable* E *is a maximal state entropy measurement with respect to any reading scale* $\mathcal{R}$.

As a further illustration of the maximal state entropy measurements we shall now consider unitary premeasurements $\mathcal{M}_U^m$ of an ordinary discrete observable A. Assuming that the eigenvalues of the pointer observable form a closed set there is a natural finest reading scale consisting of the pointer eigenvalues. We shall assume throughout the present subsection that this reading scale is chosen, so that the letter $\mathcal{R}$ may occasionally be dropped. If φ is an initial vector state of $\mathcal{S}$ then $U(\varphi \otimes \Phi) = \sum N_i \gamma_i \otimes \Phi_i$ is the final state of $\mathcal{S} + \mathcal{A}$ (Section 2.3). It follows immediately that $\mathcal{M}_U^m$ is a maximal state entropy measurement if and only if the vectors γ_i are mutually orthogonal for each φ. According to Theorem 3.2.1 this is the case exactly when $\mathcal{M}_U^m$ is a strong state-correlation measurement. Note that this does not require $\mathcal{M}_U^m$ to be a repeatable nor a strong value-correlation measurement (see Theorems 3.4.1 and 3.6.2). Hence we have the following result.

THEOREM 4.1.2. *Let* $\mathcal{M}_U^m$ *be a premeasurement of an ordinary discrete observable* A *with* $U(\varphi_{ij} \otimes \Phi) = \psi_{ij} \otimes \Phi_i$. $\mathcal{M}_U^m$ *is a maximal state entropy measurement if and only if the set* $\{\psi_{ij}\}$ *is orthonormal.*

The above statements apply to the von Neumann and Lüders measurements of A, which are repeatable and therefore also maximal state entropy measurements. With a notation similar to that of Section 2.3, if φ is an initial vector state of $\mathcal{S}$, then the final states of $\mathcal{S}$ after a Lüders measurement and a von Neumann measurement of A are

$$(8) \qquad\qquad T_L(A, \varphi) = \sum N_i^2 P[\varphi_{a_i}]$$

$$(9) \qquad\qquad T_{vN}(A, \varphi) = \sum N_i^2 T_{vN,i}$$

Here $T_{vN,i} = N_i^{-2} \sum_j P[\varphi_{ij}] P[\varphi] P[\varphi_{ij}]$ if $N_i^2 \neq 0$, and $T_{vN,i} = 0$ otherwise. We consider only those von Neumann measurements of A which are induced by a maximal refinement of A. Application of Equation (7) gives

(10) $$S(T_L(A, \varphi)) = S(\{N_i^2\})$$
(11) $$S(T_{vN}(A, \varphi)) = \sum N_i^2 S(T_{vN,i}) + S(\{N_i^2\})$$

where now the numbers $S(T_{vN,i})$ depend on φ. If the degeneracy of an eigenvalue a_i is finite and equal to $n(i)$, say, then $S(T_{vN,i}) \leq ln(n(i))$.

In a measurement $\mathcal{M}$, with a reading scale $\mathcal{R}$, the change of state entropy $S(T) \mapsto S(T_\Omega)$ can be used in various ways to characterize $\mathcal{M}$. Above we studied only the entropy of the final state T_Ω with respect to a given reading scale $\mathcal{R}$. In addition, one may ask under what conditions the difference

$$S(T_\Omega) - S(T)$$

would be nonnegative. Clearly this is the case whenever the initial state is a vector state. We note also that for a Lüders measurement of A we have $S(T_L(A, T)) - S(T) \geq 0$ for any T [130]. Another result is the Groenewold-Lindblad-Ozawa inequality [89,129,158] which shows under what conditions the quantity

$$\Delta \doteq S(T) - \sum N_i^2 S(T_{\Omega,i})$$
$$= [S(T_\Omega) - \sum N_i^2 S(T_{\Omega,i})] - [S(T_\Omega) - S(T)]$$

is nonnegative. The quantity Δ was considered by Groenewold (1971) as the difference between the potential information gain upon reading the result and the increase in the deficiency of information about the actual state of the system. However, in the general context of measurement theory such an interpretation of this quantity is not possible. It presupposes repeatable measurements (cf. the next subsection). The domain of validity of the Groenewold-Lindblad-Ozawa inequality can be easily illustrated by the unitary premeasurements $\mathcal{M}_U^m$ of discrete ordinary observables A. Such measurements share the property that any final component state is a vector state $(P[\gamma_i])$ whenever the initial state T is such $(T = P[\varphi])$. This property is indeed characteristic of the nonnegativity of the above quantity [158]. Hence in the case of a $\mathcal{M}_U^m$-measurement of A we have, in fact, the trivial result: $S(T) - \sum N_i^2 T_{\Omega,i}$ $= 0 - 0 = 0$ whenever $T = P[\varphi]$.

A further option of characterizing $\mathcal{M}$ via the entropy change is to study the relative state entropy

$$S(T|T_\Omega) \doteq tr\left[Tln(T) - Tln(T_\Omega)\right].$$

This quantity is finite only if the range of T is contained in the range of T_Ω [130]. Thus, when applied to a unitary measurement $\mathcal{M}_U^m$ of a discrete ordinary observable A, the relative state entropy

$$S\left(P[\varphi]|T_\Omega\right) = -\langle\varphi|ln\left(\sum N_i^2 P[\gamma_i]\right)\varphi\rangle$$

is finite for each φ if and only if $\mathcal{M}_U^m$ is a Lüders measurement. In this case one has $S\left(P[\varphi]|T_L\right) = S(\{N_i^2\})$.

4.2 The concept of information.

The notion of average information of a probability measure, as introduced by Shannon, can be employed to yield characterizations of measurements. Consider a measurement $\mathcal{M}$ of an observable E with a given reading scale $\mathcal{R}$. Then for any state T the *average deficiency of information* of the discrete probability measure

$$\{N_i^2\} = \left\{E_T(X_i) \ : \ X_i \in \mathcal{R}\right\}$$

is given by the quantity

(12) $$H(E,T;\mathcal{R}) \doteq S(\{N_i^2\}).$$

In accordance with the minimal interpretation $H(E,T;\mathcal{R})$ can be interpreted as the (average) *deficiency of information for predicting a measurement outcome* X_i in the initial state T of the system $\mathcal{S}$. In particular, the case $H(E,T;\mathcal{R}) = 0$ of vanishing deficiency of information occurs exactly when T is an eigenstate of E, that is, if $E(X_i)T = T$ holds for some $X_i \in \mathcal{R}$.

Similarly, for the state T_Ω of $\mathcal{S}$ after the measurement one has

(13) $$H(E, T_\Omega; \mathcal{R}) = S(\{E_{T_\Omega}(X_i)\})$$

Assuming that value objectification takes place in the E-measurement $\mathcal{M}$, then this number could be interpreted as the *potential information gain* upon reading the result $X_i \in \mathcal{R}$. On this assumption it becomes interesting to compare the initial deficiency of information, given by $H(E, T; \mathcal{R})$, with the potential information gain, $H(E, T_\Omega; \mathcal{R})$ and to ask under what conditions they are the same.

Since $T_\Omega = \sum N_i^2 T_{\Omega,i}$ and hence $E_{T_\Omega} = \sum N_i^2 E_{T_{\Omega,i}}$, we have the basic inequalities [130]

(14) $\sum N_i^2 H(E, T_{\Omega,i}; \mathcal{R}) \le H(E, T_\Omega; \mathcal{R})$

$$\le \sum N_i^2 H(E, T_{\Omega,i}; \mathcal{R}) + H(E, T; \mathcal{R})$$

which provide a means of comparing the quantities $H(E, T; \mathcal{R})$ and $H(E, T_\Omega; \mathcal{R})$. If $\mathcal{M}$ is of the first kind, that is $E_T = E_{T_\Omega}$, then we have

(15) $$H(E, T; \mathcal{R}) = H(E, T_\Omega; \mathcal{R})$$

for any initial state T of $\mathcal{S}$ and for any reading scale $\mathcal{R}$. The quantity $\sum N_i^2 H(E, T_{\Omega,i}; \mathcal{R})$, which appears in the inequalities (14), describes the average deficiency of information on the value of E when the system is known to be in one of the states $T_{\Omega,i}$, $X_i \in \mathcal{R}$, after the measurement. Equality (14) does not require the vanishing of this quantity, as exemplified by the measurement defined via Equation (3.11). However, for repeatable measurements this quantity is always zero:

(16) $$H(E, T_{\Omega,i}; \mathcal{R}) = 0,$$

for all $i = 1, 2, \cdots$, for any reading scale $\mathcal{R}$. On the other hand, $H(E, T_{\Omega,i}; \mathcal{R}) = 0$ only if $E(X_j)T_{\Omega,i} = T_{\Omega,i}$ for some $X_j \in \mathcal{R}$. Hence if we require that (16) holds for each initial state T and for any reading scale $\mathcal{R}$, then $\mathcal{M}$ necessarily leaves the object system in a mixture of eigenstates of the measured observable. This does not imply the repeatability of $\mathcal{M}$. For example, a unitarily transformed Lüders measurement, with $\mathcal{I}_{\mathcal{M}}(\{\omega_i\})T = UE_iTE_iU^{-1}$, of a discrete nondegenerate ordinary observable E, with an U permuting the eigenvectors of E, satisfies conditions (15) and (16), though it need not be repeatable.

Although repeatable measurements are not uniquely singled out by these requirements, they nevertheless are optimal in the sense that for any initial state of the object system and for any given reading scale the deficiency of information for predicting a measurement outcome always equals the potential information gain upon reading the result, and there is no remaining deficiency of information on the value of the measured observable in the final component states of the object system. We summarize this result in the following theorem.

THEOREM 4.2.1. *In a repeatable measurement of any observable the deficiency of information for predicting a measurement outcome always equals the potential information gain upon reading the measurement result.*

Repeatable measurements are known to possess the following important properties: they are maximal state entropy measurements, and for them the deficiency of information always equals the potential information gain. These two aspects allow one to establish a quantitative connection between the entropy of the final state (3) and the average informations (12) and (13). Indeed we have

$$
(17) \qquad S(T_\Omega) = \sum_i N_i^2 S(T_{\Omega,i}) + H(E,T;\mathcal{R})
$$
$$
= \sum_i N_i^2 S(T_{\Omega,i}) + H(E,T_\Omega;\mathcal{R})
$$

In particular, this yields the following inequality

$$
(18) \qquad S(T_\Omega) - H(E,T;\mathcal{R}) = S(T_\Omega) - H(E,T_\Omega;\mathcal{R})
$$
$$
= \sum_i N_i^2 S(T_{\Omega,i}) \geq 0
$$

Hence for repeatable measurements the deficiency of information $S(T_\Omega)$ about the actual final state of the system is never less than the potential information gain $H(E,T_\Omega;\mathcal{R})$ upon reading the actual result. The quantities coincide exactly when all component states $T_{\Omega,i}$ are vector states. For arbitrary repeatable measurements this need not be so.

We can illustrate the above results as characterizations of the von Neumann and Lüders measurements of a discrete ordinary observable A. Since these measurements are repeatable, the relations (17) may be applied. For a Lüders measurement of A one obtains

$$(19) \qquad S(T_L(A,\varphi)) = H(A, P[\varphi]) = H(A, T_L(A,\varphi))$$

for any initial vector states φ. On the other hand, for a (maximal) von Neumann measurement of A one has only

$$(20) \ S\big(T_{vN}(A,\varphi)\big) - H\big(A, P[\varphi]\big) = S\big(T_{vN}(A,\varphi)\big) - H\big(A, T_{vN}(A,\varphi)\big)$$
$$= \sum N_i^2 S(T_{vN,i}) \geq 0$$

so that the deficiency of information on the actual state of the system after the measurement is greater than the potential information gain upon reading the result.

As in the case of state entropy, there are several alternative information theoretical characterizations of the probability changes $E_T \mapsto E_{T_\Omega}$ associated with an E-measurement $\mathcal{M}$. In addition to the above considerations we may mention the concept of average relative information, which can be used to compare the initial and final probability measures E_T and E_{T_Ω}. Assume that the final probability measure E_{T_Ω} is absolutely continuous with respect to E_T, that is, $E_{T_\Omega}(X) = 0$ whenever $E_T(X) = 0$. Let f be the Radon-Nikodym derivative of E_{T_Ω} with respect to E_T [94]. Then the *average relative information* of E_{T_Ω} *with respect to* E_T is defined as the nonnegative number [108]:

$$(21) \qquad H(E_{T_\Omega}|E_T) = \int f \, ln(f) \, dE_T$$

Consider an E-measurement $\mathcal{M}$ with a given reading scale $\mathcal{R}$. Assuming that the final probability measure $\{E_{T_\Omega}(X_i)\} \equiv \{q_i\}$ is absolutely continuous with respect to the initial one $\{E_T(X_i)\} \equiv \{p_i\}$ we get

$$(22) \quad H\big(\{E_{T_\Omega}(X_i)\} \,|\, \{E_T(X_i)\}\big) = \sum \frac{q_i}{p_i} ln\big(\frac{q_i}{p_i}\big) p_i$$
$$= -\sum q_i ln(p_i) - H(E, T_\Omega; \mathcal{R})$$

This relation can be used, in particular, for a characterization of repeatable unitary measurements $\mathcal{M}_U^m$ of A. Assuming that the final probability measure is absolutely continuous with respect to the initial probability measure for any initial (vector) state of $\mathcal{S}$, it follows that the $\mathcal{M}_U^m$-generating set $\{\psi_{ij}\}$ (cf. Section 2.3) is a complete set of eigenvectors of A. Hence $\mathcal{M}_U^m$ is repeatable so that one has $H(E_{T_\Omega}^A|E_T^A) = 0$ for any state T.

The notion of average information (12) of a discrete probability measure is not the only one ever proposed in the context of quantum mechanics. Let A be a discrete ordinary observable such that the degree of degeneracy $n(i)$ of any eigenvalue a_i is finite. Then the formula

$$(23) \qquad H^*(A,T) \doteq -\sum_i E_T^A(\{a_i\}) \, ln\left(\frac{E_T^A(\{a_i\})}{n(i)}\right)$$

defines an information functional considered by Everett (1957). Obviously we have

$$(24) \qquad H^*(A,T) = H(A,T) + \sum_i E_T^A(\{a_i\}) \, ln(n(i)) \geq H(A,T).$$

In order to give a first intuitive comparison of the two concepts of information H and H^*, we shall consider a situation in which the outcome a_k is known before the measurement, that is, $E_T^A(\{a_i\}) = \delta_{ik}$. It follows that $H(A,T) = 0$ so that there is no ignorance left about the measuring result. On the other hand, we find $H^*(A,T) = ln(n(k))$, that is, there is still some deficiency of information. This result can be explained in the following way. Even if the result a_k is known, nothing is known about the pure state φ_{kj} of $\mathcal{S}$ which lies in the subspace with dimension $n(k)$. There are $n(k)$ orthogonal states of this kind with probability $N_{kj}^2 = \frac{1}{n(k)}$, that is, the system with eigenvalue a_k is described by the state $T_k = \sum_{j=1}^{n(k)} \frac{1}{n(k)} P[\varphi_{kj}]$. In order to determine the pure state φ_{kj} of $\mathcal{S}$, one could measure some nondegenerate observable $B_k = \sum_{j=1}^{n(k)} b_j P[\varphi_{kj}]$. The deficiency of information about the value b_j

and thus about the state φ_{kj} is then given by

$$(25) \qquad H(B_k, T_k) = -\sum_j N_{kj}^2 \ln N_{kj}^2$$

$$= -\sum_j \frac{1}{n(k)} \ln\left(\frac{1}{n(k)}\right) = \ln(n(k))$$

in accordance with the formula for $H^*(A, T)$. Hence $H^*(A, T)$ describes not only the deficiency of information about the value a_k but also about the pure state of the system S.

This consideration leads to the following more general remark. Let A, B be two ordinary discrete observables such that B is a refinement of A, so that $A = f(B)$ for some function f. It follows that the potential information gain with respect to B is larger than that for A:

$$(26) \qquad H(B, T) \geq H(f(B), T).$$

For a given state T, there exists a maximal (nondegenerate) refinement $B = A_0 = \sum_{ij} a_{ij} P[\varphi_{ij}]$ of A such that the probabilities $N_{ij}^2 \doteq E_T^{A_0}(\{a_{ij}\})$ are independent of the degeneracy index j, that is, $N_{ij}^2 = \frac{1}{n(j)} E_T^A(\{a_i\})$. This T-dependent choice of A_0 leads to

$$(27) \qquad H(A_0, T) = H^*(A, T)$$

which again illustrates the meaning of the entity $H^*(A, T)$. Furthermore, the inequality (24) is seen to be a special case of (26).

4.3 Information and commutativity.

The commutativity of two bounded ordinary discrete observables A and B can be given an operational meaning by means of Lüders measurements. If T is an initial state of S, then $T_L(A, T)$ is the state of S after a Lüders measurement of A. For a measurement of another discrete ordinary observable B we may (α) measure B directly on the system S in the state T, or (β) first perform a Lüders premeasurement of A on the system in the state T and then perform a B-measurement on S in the state $T_L(A, T)$. Even if the operators A and B commute, the outcome of a single B-measurement will in general be different in the two cases (α) and (β) since in the case (β) the A-measurement produces changes of the state of the system, which may influence the result of the succeeding B-measurement. However, in the statistical average these differences disappear. This is the content of the following theorem due to Lüders (1951).

THEOREM 4.3.1. *Bounded discrete ordinary observables A and B commute if and only if $tr[TB] = tr[T_L(A,T)B]$ for all states T.*

This theorem provides an operational meaning of the concept of commutativity: if for any preparation T a Lüders measurement (without reading) of A has no influence on the expectation value of B, then it follows that A and B commute, and vice versa. It is essential for the theorem that only Lüders measurements of the discrete observable A are considered. However, it is straightforward to generalize the theorem for continuous observables in the sense of referring to all possible discretized versions of them. The intimate connection between Lüders measurements and the concept of information makes it possible to characterize the commutativity also in terms of information theory. The deficiencies of information about B which correspond to the measuring procedures (α) and (β) are $H(B,T)$ and $H(B,T_L(A,T))$ respectively. An information theoretical characterization of the commutativity can then be formulated in the following way [216]:

THEOREM 4.3.2. *Discrete ordinary observables A and B commute if and only if $H(B,P[\varphi]) = H(B,T_L(A,\varphi))$ for all vector states φ.*

This means that A and B commute if and only if for any state $P[\varphi]$ the deficiency of information about B does not depend on whether a Lüders measurement (without reading) of A has been performed or not. A Lüders measurement (without reading) of an observable A which does not commute with B may thus change the deficiency of information about B. It is remarkable that the initial ignorance $H(B,\varphi)$ about B can be either increased or decreased in this way [216].

5. Objectification

5.1 The problem.

A premeasurement of an observable is a measurement if it satisfies the *objectification* requirement. This condition rests on the idea that a measurement leads to a definite result. Objectification has proved to be the key problem of the quantum theory of measurement.

If a premeasurement $\mathcal{M}$ of an observable E is performed on the object system $\mathcal{S}$ in the state T then it leads to the final state $V(T \otimes T_A)$ of $\mathcal{S} + \mathcal{A}$. The final states of $\mathcal{S}$ and $\mathcal{A}$ are the reduced states

$$(1) \qquad T_\Omega \doteq \mathcal{I}_\mathcal{M}(\Omega)T = \mathcal{R}_\mathcal{S}\big(V(T \otimes T_A)\big)$$

$$(2) \qquad T_{A,\Omega_A} \doteq \mathcal{R}_A\big(V(T \otimes T_A)\big)$$

In order to formulate the idea of definite measurement results we refer to the notion of objectivity of an observable (Section II.2.3). In fact *pointer objectification* has been obtained exactly when the pointer observable P_A is objective in the final apparatus state (2). If, in addition, the measured observable E is objective in the final object state (1), then also *value objectification* has been obtained. Since, from the outset, the pointer observable need not be a discrete observable and the pointer function f need not be an identity function, some care is required in the elaboration of these conditions.

Let $\mathcal{R}$ be a reading scale of the pointer observable P_A, that is, a partition of the pointer value space $\Omega_A = \cup f^{-1}(X_i)$ induced by a partition of the value space of the measured observable, $\Omega = \cup X_i$, $X_i \in \mathcal{F}$, $X_i \cap X_j = \emptyset$ for $i \neq j$. Registration and reading of a result will be considered with respect to a given reading scale. A reading scale $\mathcal{R}$ determines a discrete, coarse-grained version of P_A,

$$(3) \qquad P_A^\mathcal{R} : i \mapsto P_{A,i} \doteq P_A\big(f^{-1}(X_i)\big).$$

The objectivity of $P_A^\mathcal{R}$ in the state T_{A,Ω_A} requires that this state is equivalent to the state:

$$(4) \qquad T_{A,\mathcal{R}} = \sum_i N_i^2 T_{A,i}$$

with $N_i^2 = tr[T_{\mathcal{A},\Omega_{\mathcal{A}}}P_{\mathcal{A},i}] = E_T(X_i)$ and $T_{\mathcal{A},i} = N_i^{-2}P_{\mathcal{A},i}^{1/2}T_{\mathcal{A},\Omega_{\mathcal{A}}}P_{\mathcal{A},i}^{1/2}$. Here, again, we have adopted the convention that $T_{\mathcal{A},i} = O$ whenever $N_i^2 = 0$. (We remark that the state (4) is the unique state, determined by (2) and the reading scale $\mathcal{R}$, which defines a generalized conditional probability [15,43]. While for the object system $\mathcal{S}$ one always has $T_\Omega = \sum \mathcal{I}_{\mathcal{M}}(X_i)T = \sum N_i^2 T_{\Omega,i}$, the apparatus states $T_{\mathcal{A},\Omega_{\mathcal{A}}}$ and $T_{\mathcal{A},\mathcal{R}}$ need not be the same. We note that in the special case of a unitary pre-measurement $\mathcal{M}_U^m$ of an ordinary discrete observable the equality of the states $T_{\mathcal{A},\Omega_{\mathcal{A}}}$ and $T_{\mathcal{A},\mathcal{R}}$, for all $T \in \mathcal{T}(\mathcal{H})_1^+$, is guaranteed exactly when $\mathcal{M}_U^m$ is a strong state-correlation measurement. Another necessary objectivity condition is that all the effects $P_{\mathcal{A},i}$ must have eigenvalue one and $tr[T_{\mathcal{A},i}P_{\mathcal{A},i}] = 1$ for all $i = 1, 2, \cdots$, whenever $N_i^2 \neq O$. Hence the objectivity of $P_{\mathcal{A}}^{\mathcal{R}}$ is achieved if and only if (i) the states (2) and (4) are equivalent, (ii) $P_{\mathcal{A},i}$ is real in $T_{\mathcal{A},i}$, and (iii) the ignorance interpretation can be applied to the particular decomposition given in (4). If this is the case, then the "actual" value $f^{-1}(X_k)$, say, of $P_{\mathcal{A}}$ can be determined without changing the apparatus state in any way. Due to the reality of $P_{\mathcal{A},i}$ in $T_{\mathcal{A},i}$ the decomposition (4) of $T_{\mathcal{A},\mathcal{R}}$ is an orthogonal decomposition. Then the applicability of the ignorance interpretation to (4) also requires that the states $V(T \otimes T_A)$ and $\sum I \otimes P_{\mathcal{A},i}^{1/2}V(T \otimes T_A)I \otimes P_{\mathcal{A},i}^{1/2}$ are equivalent (cf. Section II.2.5).

We conclude that the equivalence

$$(5) \qquad V(T \otimes T_A) \cong \sum I \otimes P_{\mathcal{A},i}^{1/2}V(T \otimes T_A)I \otimes P_{\mathcal{A},i}^{1/2}$$

is a necessary condition for the objectivity of $P_{\mathcal{A}}^{\mathcal{R}}$ in the final apparatus state $T_{\mathcal{A},\Omega_{\mathcal{A}}}$. It is also a necessary condition for the objectivity of $I \otimes P_{\mathcal{A}}^{\mathcal{R}}$ in the final state $V(T \otimes T_A)$ of $\mathcal{S} + \mathcal{A}$. If this condition holds for all possible initial states of the object system $\mathcal{S}$, then the *pointer objectification with respect to a reading scale* $\mathcal{R}$ has been achieved. Obviously, condition (5) is not satisfied in general. A *sufficient* condition for (5) is that $P_{\mathcal{A}}^{\mathcal{R}}$ is a classical observable of $\mathcal{A}$ (that is, each $P_{\mathcal{A},i}$ commutes with all observables of $\mathcal{A}$). In the important case of a unitary measurement $\mathcal{M}_U^m$ of a discrete ordinary observable the classical nature of the pointer observable will be seen to be even *necessary* for (5) (provided that the object system is a proper quantum system).

If the pointer objectification with respect to $\mathcal{R}$ had been obtained, one would still face the question of whether the registered value $f^{-1}(X_k)$ of P_A pertains to the system $\mathcal{S}$ as a real property. First of all one could consider the requirement that the premeasurement $\mathcal{M}$ would produce strong correlations between the final component states $T_{\Omega,i}$ and $T_{A,i}$ of $\mathcal{S}$ and $\mathcal{A}$. If these component states were vector states, then the state-correlation conditions $\rho(T_{\Omega,i}, T_{A,i}, V(T \otimes T_A)) = 1$ for all $i = 1, 2, \cdots$ and for all T (for which $0 \neq N_i^2 \neq 1$) are realized exactly when $T_{A,\Omega_A} = T_{A,\mathcal{R}}$. Yet it could happen that $tr[T_{\Omega,i}E(X_i)] \neq 1$, so that $E(X_i)$ is not real in the state $T_{\Omega,i}$. Thus to allow for the objectivity of E, one could stipulate further that the premeasurement $\mathcal{M}$ would produce also strong value-correlations: $\rho(E(X_i), P_{A,i}, V(T \otimes T_A)) = 1$ for all $i = 1, 2, \cdots$ and for all T (for which $0 \neq N_i^2 \neq 1$). In the case of an ordinary observable E strong value-correlation is achieved whenever the measurement is of the first kind. Then the relation $tr[T_{\Omega,i}E(X_i)] = 1$ holds for each $i = 1, 2 \cdots$, and for any initial state T of $\mathcal{S}$ (with $tr[TE(X_i)] \neq 0$). Under these conditions the measurement $\mathcal{M}$ leads to *value objectification with respect to the reading scale $\mathcal{R}$.*

5.2 Classical pointer observable.

We now illustrate the above approach in the case of a normal unitary premeasurement $\mathcal{M}_U^m$ of an ordinary discrete observable A. If φ is an initial (vector) state of $\mathcal{S}$, then the final states of $\mathcal{S} + \mathcal{A}$, $\mathcal{S}$, and $\mathcal{A}$ are

$$
(6) \qquad U(\varphi \otimes \Phi) = \sum c_{ij}\psi_{ij} \otimes \Phi_i = \sum N_i \gamma_i \otimes \Phi_i
$$

$$
(7) \qquad T_U = \sum N_i^2 P[\gamma_i]
$$

$$
(8) \qquad T_{A,U} = \sum N_i N_k \langle \gamma_i | \gamma_k \rangle |\Phi_k\rangle\langle\Phi_i|
$$

respectively. A natural reading scale now is $\mathcal{R} = \cup\{a_i\}$, so that $A_A = A_A^{\mathcal{R}}$, and the state of $\mathcal{A}$ defined by $\mathcal{R}$ and $T_{A,U}$ is

$$
(9) \qquad T_{A,\mathcal{R}} = \sum N_i^2 P[\Phi_i].
$$

It will be immediately observed that the states (8) and (9) are the same for all φ if and only if the U-defining set $\{\psi_{ij}\}$ is orthonormal. In general this is not the case. The necessary condition (5) for the pointer objectification (with respect to $\mathcal{R} = \cup\{a_i\}$) now reads

$$(10) \qquad P[U(\varphi \otimes \Phi)] \cong \sum I \otimes P[\Phi_i] P[U(\varphi \otimes \Phi)] I \otimes P[\Phi_i].$$

Let $R \in \mathcal{P}(\mathcal{H}_\mathcal{A})$. If one requires (10) to hold for a proper quantum system, that is,

$$\langle U(\varphi \otimes \Phi)|P \otimes RU(\varphi \otimes \Phi)\rangle = \sum \langle U(\varphi \otimes \Phi)|P \otimes P[\Phi_i]RP[\Phi_i]U(\varphi \otimes \Phi)\rangle$$

for all $P \in \mathcal{P}(\mathcal{H}_\mathcal{S})$ and for all $\varphi \in \mathcal{H}_\mathcal{S}$, then it follows that R must commute with all $P[\Phi_i]$. Obviously the commutativity of R with $A_\mathcal{A}$ is also sufficient for (10). This shows that, in the present approach, the commutativity of the pointer observable with any other observable of the apparatus is a *necessary and sufficient* condition for the pointer objectification. With this we have established the following result.

THEOREM 5.2.1. *Let $\mathcal{M}_U^m$ be a normal unitary premeasurement of an observable A performed on a proper quantum system $\mathcal{S}$. Then pointer objectification is obtained if and only if the pointer observable $A_\mathcal{A}$ is a classical observable.*

In the case of a classical maximal pointer observable $A_\mathcal{A}$ any other observable of $\mathcal{A}$ is a function of $A_\mathcal{A}$. It follows further that there is no longer a one-to-one correspondence between the ordinary observables and the self-adjoint operators in $\mathcal{H}_\mathcal{A}$. Consequently a state T of $\mathcal{A}$ corresponds to an equivalence class $\{T' : T' \cong_{A_\mathcal{A}} T\}$ of $\mathcal{T}(\mathcal{H}_\mathcal{A})_1^+$ and the presence of a one-dimensional projection operator in this equivalence class does not imply that the state is pure. We remind ourselves that a vector state is pure if and only if its equivalence class contains only one element (see, for instance, Refs. 15, 107 or 168).

The above theorem can be extended to more general measurement situations, as, for example, to an A-measurement induced by a measurement $\mathcal{M}_U^m$ of a refinement B of $A = f(B)$. The equivalence (5) still assumes the form (10), but now it only follows that the degenerate pointer observable $f(A_\mathcal{A})$ is a classical observable of $\mathcal{A}$. In any case the objectification requirement leads to the conclusion that the apparatus must have some classical properties, excluding, in particular, superpositions of pointer states corresponding to different readings [33].

We assume now that the pointer observable A_A of the premeasurement $\mathcal{M}_U^m$ of A is, indeed, classical. In that case *the only pure states of the measuring apparatus $\mathcal{A}$ are the eigenstates $P[\Phi_k]$ of the pointer observable*. The pointer objectification condition (10) is fulfilled. In particular, the final apparatus state (8) is equivalent to (9). The decomposition (9) is the only decomposition of $T_{A,R}$ into pure states of $\mathcal{A}$. This means, in particular, that the pointer probabilities N_i^2 as well as the final state of $\mathcal{A}$ allow an ignorance interpretation: when the apparatus $\mathcal{A}$ is in the state $T_{A,R}$, then it is actually in one of the pure states $P[\Phi_k]$, the coefficients N_i^2 describing our knowledge about the actual state of $\mathcal{A}$. The *actual value* of the pointer observable A_A can be read without changing the actual state of $\mathcal{A}$.

The final state of the object system $\mathcal{S}$ is not directly affected by the assumption that A_A is classical. However, if the measurement $\mathcal{M}_U^m$ produces strong correlations between the component states $P[\gamma_i]$ and $P[\Phi_i]$ (in which case the γ_i are pairwise orthogonal), then the ignorance interpretation can be applied to T_U, as well. Assuming that this is the case, if $P[\Phi_k]$ is the actual final state of $\mathcal{A}$, then $P[\gamma_k]$ is the actual final state of $\mathcal{S}$. Then $\langle\gamma_k|E^A(\{a_k\})\gamma_k\rangle$ is the probability that in the actual final state of $\mathcal{S}$ the measured observable A has the value a_k. This probability need not be 1. If , in addition, the measurement $\mathcal{M}_U^m$ is a strong value-correlation (or first kind) measurement, then the vectors ψ_{ij} are eigenvectors of A, and A is objective in the actual final state of $\mathcal{S}$. The value objectification is thereby achieved.

In the present approach the classical nature of the pointer observable in a measurement $\mathcal{M}_U^m$ of a discrete ordinary observable is a sufficient (as in the general case) but also a necessary condition for the pointer objectification. The problem of how to realize classical (pointer) observables within quantum mechanics is not discussed here. In fact we are unable to solve this problem. Nevertheless the above result shows what is needed for ensuring the consistency of the minimal interpretation with the assumption that quantum mechanics is a complete theory of individual objects.

In addition to the question of how to explain the existence of a classical pointer observable, this solution of the objectification problem bears with itself some further problems. Firstly the assumption that the unitary measurement coupling U represents an observable H of $S + A$ via the relation $U = e^{iH}$ cannot be reconciled with the classical nature of A_A. Since on the other hand the classical nature of A_A is inevitable for the pointer objectification and thus for the measurement, one arrives at the surprising conclusion that the unitary operator U represents a measuring coupling only if H is not an observable (Section 6.2). Secondly, if A_A is a discrete maximal observable, then its classical nature forces the apparatus A to be a discrete classical system. But such a system cannot be a carrier of the Galilei covariant canonical position and momentum observables [140,168]. Generally a quantum system does represent a Galilean system of imprimitivity in a separable Hilbert space. It follows that *a measuring apparatus is not a quantum mechanical system*. Obviously this conclusion invalidates some of the presuppositions of a universally valid quantum mechanics (Section 1.2).

With the above problems arising from the fulfilment of the objectification requirement, we are facing the following important conclusions (see Table I.1). If it is possible to explain the origin (and to ensure the existence) of classical (pointer) observables within quantum mechanics, then the incorporation of fundamental symmetries (Galilei covariance) still requires a framework more general than that furnished by separable Hilbert spaces. Hence quantum mechanics cannot be a universal theory. If, on the other hand, quantum mechanics is unable to account for classical (pointer) observables, then there are two possible conclusions. Either one believes that (continuous) superselection rules do exist in a strict sense, which in turn means that one has to give up the universal validity of quantum mechanics. Or one accepts that quantum mechanics properly accounts for the fact that superselection rules and classical observables are only approximately realized in nature. Then quantum mechanics is a universal theory for a fundamentally unsharp reality.

The preceding discussion is based on the validity of the assumptions of Theorem 5.2.1. New possibilities arise if one gives up the axiom of unitary dynamics. In fact in the literature there have been "insolubility theorems" for the measurement problem similar to the one above. Regarding (tacitly) both S and A as proper quantum systems, Wigner (1963) concluded that the linearity of the quantum mechanical dynamics cannot be maintained if objectification is to be achieved. This result was corroborated and generalized by other authors [28,68,189]. These alternative approaches will be reviewed in more detail in Chapter IV.

5.3 Registration and reading.

Registration and reading are the final steps in a measuring process. First, after the measurement interaction, the apparatus reaches a stage at which it records some outcome; that is, the pointer assumes some value on the reading scale. In this sense the process of *registration* is nothing but the pointer objectification, so that the respective apparatus state $T_{A,U} \cong T_{A,R}$ admits an ignorance interpretation with respect to its components $T_{A,i}$. The remaining step, the *reading*, is performed by the observer, who in this way eliminates his ignorance and changes the description of the apparatus state according to the registered outcome.

In the preceding Section the reading scale was considered *discrete*. In fact, it was defined as a partition of the value space of the pointer observable. There exist various types of arguments indicating that this discreteness is mandatory.

a) Pragmatic need. Any physical experiment is designed to yield definite outcomes out of a collection of alternatives. These outcomes must be described by essentially finite means, either by digital recordings, or by estimating a pointer position in terms of a rational number on an apparently continuous scale.

b) Statistics requirement. The statistical evaluation of experimental outcomes is based on counting frequencies of mutually exclusive events out of a countable collection. This again requires the fixing of a partition of the value space of the pointer observable.

c) Pointer objectification. The fact that the pointer ultimately assumes a definite position is to be interpreted as a repeatable measurement of the pointer observable. Hence either this observable itself or one of its (actually measured) coarse-grained versions must be discrete.

The discrete nature of the reading scale entails that a given measuring apparatus allows only a measurement of a discrete version of the observable under consideration. Since the value objectification also requires repeatability (Section 5), it can be achieved only if the measured observable is discrete. Hence the conclusion that only discrete versions of observables are experimentally accessible can be reached from various lines of argument. With this we do not, however, deny the operational relevance of continuous observables. On the contrary, their usefulness as idealizations shows itself in the fact that they represent the possibility of indefinitely increasing the accuracy of measurements by choosing increasingly refined reading scales.

The above three arguments for the need for discrete reading scales can be substantiated in formal terms. First, the pragmatic argument $a)$ is best illustrated by means of the information concept. Let E be a continuous ordinary observable on $\mathcal{B}(\Re)$, meaning that for all $X \in \mathcal{B}(\Re)$ there exists $Y \in \mathcal{B}(\Re)$ such that $Y \subset X$ and $O \neq E(Y) < E(X)$. Further, let $\mathcal{R}_1$, $\mathcal{R}_2$, ... be a sequence of increasingly finer reading scales on $\mathcal{B}(\Re)$ such that $\mathcal{R}_{n+1}$ consists of partitions of the elements of $\mathcal{R}_n$. As shown in Section 4, we have

$$(11) \qquad\qquad H(E,T;\mathcal{R}_{n+1}) \geq H(E,T;\mathcal{R}_n)$$

for all states T and for all n. Assume that the partitions approach points so that the maximum size of the partition intervals $X_i^{(n)}$, $i = 1, 2, ...$, of $\mathcal{R}_n$ tends to zero in the following sense: for each state T, the sequence $sup\{tr\,[TE(X_i^{(n)})] : i = 1, 2, ...\}$ converges to zero. Then it follows that the sequence of numbers $H(E,T;\mathcal{R}_n)$ increases indefinitely for all T. Hence if there were a continuous reading scale then an E-measurement would have to lead to an infinite increase in information.

Argument $b)$ refers to the intended empirical content of the minimal interpretation of the probability measures E_T: if a measurement of E in a state T had been repeated n times, and the result X had occurred ν times, then $\lim_{n\to\infty} \nu/n = E_T(X)$. Two ways for a formal justification of such a statistical interpretation were discussed in Section III.2.4. Here we recall only that the measurement statistics interpretation of the probabilities $E_T(X)$, $X \in \mathcal{F}, T \in \mathcal{T}(\mathcal{H})_1^+$, with respect to a given E-measurement $\mathcal{M}$, induces a family of discretized pointer observables

P_A^R, $\mathcal{R}$ a reading scale, from which the relevant probabilities

$$(12) \qquad P_{A,\mathcal{R}_A(V(T\otimes T_A))}^{\mathcal{R}}(\{i\}) = P_{A,\mathcal{R}_A(V(T\otimes T_A))}(f^{-1}(X_i^{(\mathcal{R})}))$$
$$= E_T(X_i^{(\mathcal{R})})$$

for any $X_i^{(\mathcal{R})}$ (in $\mathcal{R}$) and for each $\mathcal{R}$ can be obtained as relative frequencies. That is, for each T and for each $\mathcal{R}$ there is a sequence $\Gamma_{T,\mathcal{R}}$ of $P_A^{\mathcal{R}}$-outcomes such that

$$(13) \qquad relf(X_i^{(\mathcal{R})}, \Gamma_{T,\mathcal{R}}) = P_{A,\mathcal{R}_A(V(T\otimes T_A))}^{\mathcal{R}}(\{i\})$$

for each i.

The third argument $c)$ referring to repeatability is based on Theorem 3.6.1. Therefore it is restricted, strictly speaking, to completely positive instruments. In fact, according to Theorem 2.2.1, an instrument admits a (normal) unitary premeasurement $\mathcal{M}_U$ exactly if it is completely positive. We do not think, however, that this is a severe limitation on the argument.

6. Measurement dynamics

6.1 The problem.

An important problem of the quantum theory of measurement related both to the possibility of premeasurements as well as to the objectification requirement is the question of the physical realizability of the appropriate measurement couplings, that is, the state transformations V in the 5-tuples $\langle \mathcal{H}_A, P_A, T_A, V, f \rangle$. Within the conventional description of the dynamics of isolated quantum systems, there are logically two possibilities: (i) the system $\mathcal{S} + \mathcal{A}$ consisting of object and apparatus can be considered as an isolated system; or (ii), the influence of the environment $\mathcal{E}$ on $\mathcal{S} + \mathcal{A}$ cannot be neglected.

In the case (i) the usual description of dynamics applies, and V should be in the range of the mapping $t \mapsto \mathcal{U}_t$, $t \in \Re$, the dynamical group of $\mathcal{S} + \mathcal{A}$ (cf. Section II.1.3.). More explicitly, $\mathcal{S}$ and $\mathcal{A}$ should be *dynamically independent* before and after the measurement, that is, before a time $t = 0$ and after some time $t = \tau > 0$. This implies that the Hamiltonian H, which generates the dynamics $\mathcal{U}_t$, $t \in \Re$, coincides with the free Hamiltonian $H_\mathcal{S} + H_\mathcal{A}$ before and after the measurement, while in the time interval $0 \leq t \leq \tau$ the measurement interaction comes into play:

$$(1) \qquad\qquad H = H_\mathcal{S} + H_\mathcal{A} + H_{int}.$$

The mapping V should be identified with $\mathcal{U}_\tau$. But then the unitary dynamics $t \mapsto U_t = exp(-\frac{i}{\hbar}Ht)$ is either discontinuous, or a time-dependent interaction $H_{int}(t)$ is to be introduced. Both possibilities are, however, excluded by the continuity and the group properties of the dynamics $t \mapsto \mathcal{U}_t$.

This problem of incorporating V into the dynamics $t \mapsto \mathcal{U}_t$ of $\mathcal{S} + \mathcal{A}$ seems to allow only solutions in the sense of some approximations. The interaction part H_{int} of the total Hamiltonian H should be negligible before and after the measuring process. If the actual duration of the interaction is of no concern, then the canonical approach is that of describing measuring processes as scattering processes, so that V is identified with the respective scattering operator [137]. However, it may be desirable to account explicitly for the finite times of preparation and registration. In this case the finite duration of the interaction, that is, the apparent time dependence of H_{int}, needs to be explained. This can be achieved by regarding the relative motion of $\mathcal{S}$ and $\mathcal{A}$ as the relevant "clock" determining approximately the times of turning on and off the interaction. In Ref. 3 it is indicated by means of a simple model how in this way an effectively time-dependent Hamiltonian is obtained from a unitary dynamical group if the system $\mathcal{A}$ is "large" in some suitable sense.

Another aspect of the problem of measurement dynamics is the limited number of interactions available. This contingent fact leads to a "natural" restriction of the set of operators which correspond to actually observable quantities. In particular, the semiboundedness of the Hamiltonian entails that the probability reproducibility may be achieved, in general, only in an approximative way [86].

Turning to the second option, case (ii) above, the measurement coupling V should result from a family $\mathcal{V}_t$, $t \in \Re$, of linear state transformations representing the reduced dynamics of $\mathcal{S} + \mathcal{A}$ derived from the unitary dynamics $t \mapsto \mathcal{U}_t$ of the isolated system $\mathcal{S} + \mathcal{A} + \mathcal{E}$. But treating $\mathcal{S} + \mathcal{A} + \mathcal{E}$ as an isolated system entails exactly the same problems as those encountered in case (i).

6.2 An inconsistency.

Besides the question of realizing the measurement coupling V as a part of the dynamics $t \mapsto \mathcal{V}_t$, the objectification requirement poses additional constraints on the measurement interactions. Apart from the fact that the measurement coupling V should give rise to suitable correlations in the final state $V(T \otimes T_\mathcal{A})$ of $\mathcal{S} + \mathcal{A}$, the apparatus should assume a definite pointer value at the end of the measurement. As shown in Section 5, this can be achieved by assuming that the pointer observable is a classical observable. This state of affairs forces one to consider a modification of quantum mechanics; in particular, it implies constraints on the dynamics of $\mathcal{A}$ as well as of $\mathcal{S} + \mathcal{A}$ [15]. The assumption that the pointer observable is classical implies a further puzzling feature, namely the measurement coupling cannot represent an observable of $\mathcal{S} + \mathcal{A}$ [16].

To discuss this problem, consider a unitary measurement $\mathcal{M}_U$ of a discrete ordinary observable A, and assume that the pointer observable $A_\mathcal{A}$ is, in fact, classical. The pointer observable $A_\mathcal{A}$ can be interpreted as an observable $I \otimes A_\mathcal{A}$ of the compound system $\mathcal{S} + \mathcal{A}$. As the von Neumann algebra $\mathcal{L}(\mathcal{H}_\mathcal{S} \otimes \mathcal{H}_\mathcal{A})$ of bounded operators on $\mathcal{H}_\mathcal{S} \otimes \mathcal{H}_\mathcal{A}$ is generated by the operators of the product form $B \otimes C$, $B \in \mathcal{L}(\mathcal{H}_\mathcal{S})$, $C \in \mathcal{L}(\mathcal{H}_\mathcal{A})$, one recognizes that $I \otimes A_\mathcal{A}$ is a classical observable of $\mathcal{S} + \mathcal{A}$ as well [193]. This fact has an important consequence. The unitary measurement coupling U can be written as $U = exp(\imath H)$. But H cannot represent an observable of $\mathcal{S} + \mathcal{A}$, like a Hamiltonian of $\mathcal{S} + \mathcal{A}$. Indeed, if H is an observable of $\mathcal{S} + \mathcal{A}$ and $A_\mathcal{A}$ is classical, then one has

$$(2) \qquad E^{A_\mathcal{A}}_{P[\Phi]} = E^{I \otimes A_\mathcal{A}}_{P[\varphi \otimes \Phi]} = E^{I \otimes A_\mathcal{A}}_{P[U(\varphi \otimes \Phi)]} = E^{A_\mathcal{A}}_{\mathcal{R}_\mathcal{A}(P[U(\varphi \otimes \Phi)])}$$

for any unit vector $\varphi \in \mathcal{H}$, since $I \otimes A_\mathcal{A}$ commutes with U. Since $E^{A_\mathcal{A}}_{P[\Phi]}$ does not depend on the initial state of the object system, Equation

(2) is incompatible with Equation (3) of Section 2, which expresses the probability reproducibility condition. Hence we have established the following result.

THEOREM 6.2.1. *Let a 4-tuple* $\langle \mathcal{H}_{\mathcal{A}}, A_{\mathcal{A}}, \Phi, U \rangle$ *be a candidate for a normal unitary premeasurement* $\mathcal{M}_U$ *of a discrete ordinary observable* A. *If* $A_{\mathcal{A}}$ *is a classical observable and if the coupling* U *is generated by an observable of* $\mathcal{S} + \mathcal{A}$, *then* $\langle \mathcal{H}_{\mathcal{A}}, A_{\mathcal{A}}, \Phi, U \rangle$ *cannot fulfil the probability reproducibility condition.*

We conclude that if $A_{\mathcal{A}}$ is a classical pointer observable associated to a premeasurement $\mathcal{M}_U$ of A, then H cannot be an observable of $\mathcal{S} + \mathcal{A}$, and vice versa.

As shown in Section 5.2, the classical nature of the pointer observable $A_{\mathcal{A}}$ is a consequence of the objectification requirement in the case of a normal unitary premeasurement $\mathcal{M}_U^m$. Therefore such a measurement cannot be realized by means of a unitary dynamical group $\mathcal{U}_t$, the generator of which being an observable of $\mathcal{S} + \mathcal{A}$. If, on the other hand, only dynamical groups of this kind are available – as it is usually assumed – then no premeasurement $\mathcal{M}_U^m$ can serve as a measurement, since the objectification is impossible. Finally, insisting on both – classical pointer $A_{\mathcal{A}}$ *and* unitary measurement dynamics $\mathcal{U}_t$ – forces one into the strange conclusion that the generator H, the Hamiltonian of $\mathcal{S} + \mathcal{A}$, is no observable.

The problem of incorporating the measurement coupling between $\mathcal{S}$ and $\mathcal{A}$ as a part of the dynamics of $\mathcal{S} + \mathcal{A}$, or $\mathcal{S} + \mathcal{A} + \mathcal{E}$, and the inconsistency between the classical nature of the pointer observable and the physical realizability of a unitary measurement coupling lead to the consideration of *modified* descriptions of *dynamics*. It has been shown that a suitable additional term in the Schrödinger equation or in the von Neumann–Liouville equation may lead to a spontaneous pointer localization and thus perhaps to the classical nature of pointer observables [77]. Still the nonunique decomposability of mixtures may require an explicit description of dynamics as a stochastic process on the level of the Gemenge representation of states. This, in its turn, seems to require *nonlinear* state transformations [78] (Section IV.4.2).

7. Limitations on measurability

In investigating the measurement possibilities of quantum mechanics, the quantum theory of measurement also reveals the limitations on measurability. There are two types of such limitations, those implied by the theory itself, and those which may arise when the theory is supplemented by some further assumptions. The question of practical limitations in the sense of what actually can be measured in a laboratory is outside the scope of this review.

In the first group there are limitations like "only discrete observables admit repeatable measurements" and "complementary observables cannot be measured together". Thus, for example, the usual position observable does not admit a repeatable measurement. Such a result is connected with our very possibilities of constituting physical objects. The fact that, say, position and (conjugate) momentum observables cannot be measured together is a basic and well-understood feature of quantum mechanics. Yet the idea that the position - momentum uncertainty relations open a way to circumvent this limitation has, however, been more controversial and has only been carried out rather recently [30,123].

Among the second type of limitations there is the fact that only observables which commute with all conserved observables can be measured at all. For instance, if momentum conservation is a universal conservation law, then position cannot be measured at all. However, conclusions of this type presuppose that the dynamical problem of the previous section has been solved.

There are also the fundamental limitations which concern the very nature of the measuring apparatus and the measurement coupling. As became evident in Section 5, no proper quantum mechanical object can serve as a measuring apparatus. Moreover, if the pointer observable is to be classical, then the unitary measurement coupling cannot represent an observable in the sense of an interaction Hamiltonian of the object - apparatus system. Or conversely, if a unitary measurement coupling is generated by an observable interaction Hamiltonian, then the pointer observable cannot be classical. These fundamental limitations are most serious, and they indicate the directions for the possible resolutions of the objectification problem. Finally there is also a limitation on the determination of the past and the future of an object system: complete (statistical) determination of the state (before the measurement) cannot be achieved by means of repeatable measurements. This phenomenon will be explained in some detail in Section 8.

In this section we shall discuss limitations on measurability implied by quantum measurement theory. Before entering into this subject, we ought to remind ourselves of some basic positive results of the measurement theory. First of all, for each observable of the object system there do exist premeasurements. Leaving aside the objectification problem, this fact confirms the idea that physical quantities are, indeed, observables, that is, they can be measured. Furthermore, for single ordinary observables no a priori limitations on their measurement accuracies arise.

7.1 Repeatable measurements and continuous observables.

If an observable E of $\mathcal{S}$ admits a unitary measurement $\mathcal{M}_U$ which is repeatable, then E is necessarily discrete. This fact (reported in Section 3.6) has an important (though obvious) corollary.

COROLLARY 7.1.1. *No continuous observable admits a repeatable unitary measurement.*

This result causes difficulties in our understanding of the operational definition of continuous observables, among them position, momentum, and energy – observables which are most important for the concept of a particle in quantum physics. We shall illustrate these difficulties by considering the localization observable of an object residing in three dimensional Euclidean space $\Re^3$. In fact the localization observable of such an object can be decomposed into three similar parts referring to the three component spaces $\Re$. If the object has a nonzero rest mass, then its localization in $\Re$ is simply the spectral measure E^Q of the usual position observable Q:

$$(1)\ (Q\varphi)(x) = x\varphi(x) \ \text{ for a.e. } \ x \in \Re, \ \varphi \in dom(Q) \subset \mathcal{H} = \mathcal{L}^2(\Re, dx)$$
$$E^Q(X)\varphi = c_X\varphi \ \text{ for any } \ X \in \mathcal{B}(\Re)$$

where c_X is the characteristic function of the set $X \in \mathcal{B}(\Re)$ [15,107]. The spectrum of Q is the whole real line $\Re$ and it has no eigenvalues (since the space has no discrete part). Q is continuous so that it does not admit any repeatable unitary measurement. This raises the question how to define Q operationally. From the results of Section 2 we know that Q admits, in particular, unitary measurements, but none of them can be repeatable. Being unable to provide a canonical answer to the question posed, we we shall content ourselves with demonstrating that the most obvious way of defining Q operationally is in fact ruled out.

To begin with, we recall that the Borel σ-algebra $\mathcal{B}(\Re)$ of $\Re$ is generated by the closed intervals I of $\Re$. Hence also the range of E^Q, $E^Q(\mathcal{B}(\Re)) = \{E^Q(X) : X \in \mathcal{B}(\Re)\}$, is generated by the projections $E^Q(I)$ associated with such intervals. This means that the mapping $I \mapsto E^Q(I)$ (on the closed intervals) extends to the spectral measure $X \mapsto E^Q(X)$ (on $\mathcal{B}(\Re)$) [203]. Thus to define Q it suffices to define the localizations $E^Q(I)$ associated with the intervals I. Each $E^Q(I)$ can be defined, for instance, via the state transformation $T \mapsto E^Q(I)TE^Q(I)$ which corresponds to the yes-outcome of the Lüders measurement of the simple observable $c_I(Q)$. A diaphragm with a slit I is a prototype of an experimental arrangement leading to this state transformation. The natural question then is whether the mapping $I \mapsto \mathcal{I}(I)$, with $\mathcal{I}(I)T = E^Q(I)TE^Q(I)$, extends to an instrument of Q, that is, whether by varying slit I in the diaphragm one can define Q. The answer to this question is negative. Indeed, if there were an instrument $\mathcal{I}^Q$ of Q such that $\mathcal{I}^Q(I) = \mathcal{I}(I)$ for all closed intervals I, then due to the additivity of the instrument

$$(2) \qquad E^Q(I)TE^Q(I) = E^Q(I_1)TE^Q(I_1) + E^Q(I_2)TE^Q(I_2)$$

for all states T and for any partition of I into disjoint subintervals I_1 and I_2. But (2) cannot hold true, for example, for the vector states φ for which $\langle \varphi | E^Q(I_1)\varphi \rangle \neq 0 \neq \langle \varphi | E^Q(I_2)\varphi \rangle$. In fact, one can show that there is no instrument $\mathcal{I}^Q$ associated to Q for which $\mathcal{I}^Q(X) = E^Q(X)TE^Q(X)$, $T \in \mathcal{T}(\mathcal{H})_1^+$, for some $X \in \mathcal{B}(\Re)$. This result holds true for all continuous observables [34].

These considerations show that an operational definition of the localization observable in terms of ideal or repeatable measurements is impossible. The usual way out of this difficulty consists of restricting oneself to discrete versions of Q (Ref. 150; see also argument b) in Section 5.3). A disadvantage of this approach is that it destroys the translation covariance characteristic of the localization concept. Another approach is to relax strict repeatability into δ-repeatability [48]. A Q-compatible instrument $\mathcal{I}$ is δ-repeatable if for all states T and all $X \in \mathcal{B}(\Re)$ one has

$$(3) \qquad tr\left[\mathcal{I}(X_\delta)\mathcal{I}(X)T\right] \; = \; tr\left[\mathcal{I}(X)T\right]$$

Here $X_\delta = \{x \in \Re : |x - x'| < \delta \text{ for all } x' \in X\}$ denotes the closed δ-neighbourhood of X. There exist δ-repeatable, covariant, completely positive Q-compatible instruments; but still the known examples are of a somewhat artificial nature. Therefore an even more general concept of approximate repeatability has been proposed admitting more natural realizations. A Q-compatible instrument is (ϵ, δ)-repeatable if it satisfies the following for all states T and all Borel sets X:

$$(4) \qquad tr\left[\mathcal{I}(X_\delta)\mathcal{I}(X)T\right] \; \geq \; (1 - \epsilon)\, tr\left[\mathcal{I}(X)T\right]$$

Here ϵ and δ are some fixed ("small") numbers. This notion of approximate repeatability may even be applied to the operational definition of general continuous observables [37].

7.2 Complementary observables.

Consider any two ordinary observables A and B of an object system S. Any of their measurements can be combined as the sequential AB– and BA–measurements (cf. Section 3.6). The induced instruments are the composite instruments $\mathcal{I}^B \circ \mathcal{I}^A$ and $\mathcal{I}^A \circ \mathcal{I}^B$, respectively. It may happen that for some measurements of A and B the sequential AB– and BA–measurements are equivalent, that is, $\mathcal{I}^B \circ \mathcal{I}^A = \mathcal{I}^A \circ \mathcal{I}^B$. In that case we say that the sequential measurements are order independent. The existence of order independent sequential measurements for a given pair of observables A and B implies their commutativity, or, in general, their coexistence (when arbitrary observables are considered) [126]. In the case of ordinary discrete observables also the converse

holds true, that is, from commutativity the existence of order independent sequential measurements is obtained. This fact is closely related
to the compatibility of A and B in the sense discussed in Section 4.

Complementary observables represent an important extreme case
of noncommutative, or noncoexistent observables. Any two observables are *complementary* if the experimental arrangements which per-
mit their unambiguous (operational) definitions are mutually exclusive
[160]. This conception of complementarity of two observables A and B
lends itself readily to a formal representation as a relation between the
A- and B-compatible instruments [123,124]. For the present purposes
it is, however, unnecessary to go into formal details. Instead we state
the obvious result:

COROLLARY 7.2.1. *Complementary observables have no order independent sequential measurements.*

As another related no-go-theorem for the measurability of complementary observables we state the one resulting from the strong non-
coexistence (total noncommutativity) of such observables.

COROLLARY 7.2.2. *Complementary observables do not admit any joint
measurements.*

Canonically conjugate position and momentum observables are complementary. The same holds true, for example, for any two (different) spin
components of a spin-$\frac{1}{2}$ object.

Finally, we recall that according to the measurement unsharpness
interpretation of the uncertainty relations complementary observables
can be measured together if the involved measurement unsharpness is
sufficiently large [96]. A systematic measurement theoretical justification of this interpretation has been worked out; but a review of this
topic is outside the scope of this treatise (for references, cf. [35].)

7.3 Measurability and conservation laws.

We consider next limitations on measurability implied by the existence of universal conservation laws. Such limitations were discovered
by Wigner (1952). To start with, we shall first restate the result of
Wigner as elaborated by Araki and Yanase (1960), using, however, our
notations.

THEOREM 7.3.1. *Let* $\langle \mathcal{H}_A, A_A, \Phi, U_L \rangle$ *be a Lüders measurement of a discrete ordinary observable* A. *Let* $\hat{K} = K \otimes I_A + I \otimes K_A$ *be a bounded self-adjoint operator on* $\mathcal{H}_S \otimes \mathcal{H}_A$. *Assume that* $\hat{K}$ *is a constant of motion of* $S + A$ *with respect to* U_L, *that is,* $[\hat{K}, U_L] = O$. *Then also* $[K, A] = O$.

This theorem suggests two divergent interpretations:

INTERPRETATION 1. *Any discrete ordinary observable* A *admits a Lüders measurement, with a measurement coupling* U_L. *If* $\hat{K}$ *is any additive (bounded) observable of* $S + A$ *which is a constant of motion with respect to* U_L, *then* K *commutes with* A. *If* $[K, A] \neq O$, *then* $\hat{K}$ *is not a constant of motion with respect to* U_L.

INTERPRETATION 2. *Assume that* $\hat{K}$ *represents a universal conservation law, that is, it is a constant of motion with respect to all physically admissible evolutions of* $S + A$. *If* $[K, A] \neq O$, *then the Lüders measurement of* A, *with* U_L, *is not a physically realizable measurement, that is,* U_L *is not (a part of) a physically admissible evolution of* $S + A$.

Interpretation 2 follows Wigner who writes: "Only quantities which commute with all additive conserved quantities are precisely measurable". (See Ref. 210, p. 298; note that here "precisely" refers to a Lüders measurement.) To accept Wigner's interpretation of the above theorem as a limitation on the measurability of certain observables it is desirable to generalize this theorem to more general measurement couplings between S and A than the Lüders measurements. The next theorem provides such a generalization [16].

THEOREM 7.3.2. *Let* $\mathcal{M}_U^m$ *be a normal unitary premeasurement of a discrete ordinary observable* A. *Let* $\hat{K} = K \otimes I_A + I \otimes K_A$ *be a bounded self-adjoint operator on* $\mathcal{H}_S \otimes \mathcal{H}_A$. *If* $\hat{K}$ *commutes with* U, *then*

(a) $\qquad\qquad\qquad K$ *commutes with* A

provided that one of the following conditions is satisfied:

(b) $\qquad\qquad\qquad \rho(P[\gamma_i], P[\Phi_i], U(\varphi \otimes \Phi)) = 1$

for any $i = 1, \cdots, N$, and for all $\varphi \in \mathcal{H}, \| \varphi \| = 1$ for which $0 \neq N_i^2 \neq 1$;

(c) $K_\mathcal{A}$ commutes with $A_\mathcal{A}$.

The objectification requirement implies that the pointer observable $A_\mathcal{A}$ is classical. Hence condition (c) of this theorem is always satisfied. Condition (b) is the strong state-correlation condition of the measurement, which is fulfilled, for instance, in the case of a repeatable measurement.

Theorem 2 extends the scope of Theorem 1 in several respect. However, the relevance of these theorems as limitations on measurability is somewhat open due to the difficulties of realizing the measurement coupling U as part of a unitary evolution $\mathcal{U}_t$ of $\mathcal{S} + \mathcal{A}$. In particular, the classical nature of the pointer observable $A_\mathcal{A}$, which ensures condition (c) of Theorem 2, entails that the generator of the group $\mathcal{U}_t$, the Hamiltonian H of $\mathcal{S} + \mathcal{A}$, cannot be an observable of $\mathcal{S} + \mathcal{A}$. Hence this theorem pertains to a rather strange measurement situation which certainly does not belong to the domain of conventional quantum mechanics.

Nevertheless these theorems suggest that conservation laws associated with fundamental symmetries may lead to limitations on the measurability of observables. Consequently it would be desirable to investigate further extensions of Theorem 1, both for general (POV) observables [99] as well as for generalized dynamics. In particular, it is an open question whether also unbounded operators $\hat{K}$ give rise to limitations. There are some results indicating that this is the case for the important position-momentum pair, that is, the conservation of momentum excludes (unitary) measurements of (discretized) position [31,191]. On the other hand, it turns out that the limitations stated in Theorem 1 can be circumvented by means of introducing an arbitrarily small inaccuracy [6,191,212]. It has been shown that such inaccurate measurements can be described in terms of POV measures representing unsharp versions of the observables to be measured [31,40].

8. Preparation and determination of states

8.1 State preparation.

In the present treatise the possibility of preparing arbitrary states $T \in \mathcal{T}(\mathcal{H})_1^+$ was taken for granted without further analysis. Nevertheless problems associated with the concept of state were encountered in several places. In Sections II.2.3 and II.2.5 it was pointed out that the interpretation of mixed states depends on the method by which these states are prepared. The consideration of compound systems and the omnipresence of interactions (Section 1) provide evidence for the fact that objects can presumably be prepared at best in an approximate way. Finally the ignorance interpretation of mixed states turns out crucial for the objectification problem (Section 5).

It seems difficult to conceive of a general theory of the preparation of systems in quantum mechanics. Yet the quantum theory of measurement allows for a modelling of the process of preparation in terms of filters. In fact, a repeatable measurement is preparatory in the sense of producing systems with definite real properties. A *filter* is a repeatable measurement applied to a class of similar systems and combined with a selection of the systems having a particular value of the measured observable. Combining a sequence of filters associated to a complete set of commuting discrete (ordinary) observables yields a filter preparing a pure state. Equivalently a pure state may be prepared by means of a filter using a Lüders measurement of a maximal (nondegenerate) discrete observable. Clearly this all presupposes that the measurement outcomes can be objectified.

A filter based on von Neumann measurements of a degenerate discrete observable prepares mixed states, the interpretation of which depends on the particular choice of pointer observables. To illustrate this point, consider an observable $A = \sum_i a_i P_i$, with the spectral projections $P_i = \sum_j P[\varphi_{ij}]$. Let $A_0 = \sum a_{ij} P[\varphi_{ij}]$ be a maximal refinement of A, so that $A = f(A_0)$, with $f(a_{ij}) = a_i$ for each j and for all i. Let $\mathcal{M}_{U_L}^m$ be a Lüders measurement of A_0. Then $\langle \mathcal{H}_A, A_A, \Phi, U_L, f \rangle$ and $\langle \mathcal{H}_A, f(A_A), \Phi, U_L \rangle$ are two equivalent von Neumann measurements of

A, leaving the object system $\mathcal{S}$ in the component states

$$(1) \quad T_i \doteq N_i^{-2} \mathcal{I}_{vN}(\{a_i\}) P[\varphi] = N_i^{-2} \sum_j P[\varphi_{ij}] P[\varphi] P[\varphi_{ij}].$$

with $N_i^2 = \langle \varphi \mid P_i \varphi \rangle \neq 0$ (or $T_i = O$ if $N_i^2 = 0$). The interpretation of this state depends, however, on the measurement applied. Indeed in the first case, the object–apparatus state after reading a value a_k is

$$(2) \quad \sum_j I \otimes P[\Phi_{kj}] U(T \otimes T_A) U^{-1} I \otimes P[\Phi_{kj}],$$

whereas in the second case the corresponding state is

$$(3) \quad I \otimes \left(\sum_j P[\Phi_{kj}]\right) U(T \otimes T_A) U^{-1} I \otimes \left(\sum_j P[\Phi_{kj}]\right).$$

Here $T = P[\varphi]$ and $T_A = P[\Phi]$. In the first case the final state (2) is a mixture representing the Gemenge $\Gamma_k = \{(|\langle \varphi_{kj}|\varphi\rangle|^2, P[\varphi_{kj}] \otimes P[\Phi_{kj}]) : j = 1, 2, ..., n(i)\}$; for it is assumed that the maximal pointer observable A_A is objective though its actual value in the set $f^{-1}(\{a_k\})$ is ignored. In the second case the entanglement between the object system and the apparatus is not completely destroyed after the reading. In fact state (3) is generally a pure correlated state. In both cases the reduced state of the object system $\mathcal{S}$ is T_k as given in (1) since the two measurements are equivalent. Their interpretation is, however, different. In the first case the state T_k represents the Gemenge Γ_k admitting thus an ignorance interpretation. In the second case the state T_k is the reduced state of a pure correlated state and, as such, does not admit a similar interpretation.

These model considerations demonstrate the possibilities of preparing states – pure states, proper quantum mixtures as well as Gemenge states – by means of filters. These possibilities presuppose that the measuring results can be objectified and thus depend on the solution of the objectification problem. However, this problem has not yet been solved.

8.2 State determination versus state preparation.

Among the principal limitations on measurements is the following mentioned in Section 7. No measurement is capable of determining an arbitrary initial state T from a single measurement outcome. Even the statistics of a measurement generally do not lead to a unique state determination. In fact for a given observable E there are in general many states $T, T', \cdots$ yielding one and the same probability distribution $E_T = E_{T'} = \cdots$. An observable is called *statistically complete* (sometimes also informationally complete) if it separates the set of states, that is, if for any two states $T, T' \in \mathcal{T}(\mathcal{H})_1^+$,

$$(4) \qquad\qquad E_T = E_{T'} \text{ implies that } \quad T = T'.$$

It is well known that no ordinary observable is statistically complete. But there do exist statistically complete (POV) observables. These facts give rise to a new mode of complementarity in quantum mechanics – the complementarity between the determinations of the past and the future of a system [36]. Indeed one can show that a statistically complete observable E does not admit a repeatable measurement. Hence if one aims at an optimal determination of the future (preparation) of a system, then one would choose a repeatable measurement. But then the measured observable is not statistically complete. Conversely, an optimal determination of the past via measurement of a statistically complete observable excludes optimal state preparation. It turns out that these two complementary goals can be reconciled in an approximative way, using the notion of (ϵ, δ)-repeatability (cf. Section 7).

Résumé – Open questions

In the light of a realistic interpretation of quantum mechanics as a theory of individual systems, the task of the quantum theory of measurement is to provide a description and an understanding of measuring processes as physical processes subject to quantum mechanics. It was seen that the main problem of measurement theory is a dynamical explanation of the objectification, that is, of the occurrence of definite measurement outcomes in view of the (general) nonobjectivity of observables before their measurement. First of all, taking for granted that the measurement coupling U is established via the unitary dynamical evolution $\mathcal{U}_t$ generated by the Hamiltonian H of $\mathcal{S} + \mathcal{A}$, the objectification requirement implies that the apparatus $\mathcal{A}$ must have some classical properties associated with the pointer positions. Since the apparatus $\mathcal{A}$ is considered as a quantum mechanical system, the classical nature of the pointer observable requires the presence of some superselection rules. Hence the major problem of the quantum theory of measurement is the explanation of the occurrence of effective superselection rules, a problem which may lead to the conclusion that quantum mechanics needs to be supplemented by a more general theory. Possible directions of generalizations are either a structure more general than that of a separable Hilbert space, or generalized dynamics. A similar conclusion must be drawn if the classical pointer observable $A_\mathcal{A}$ is considered to be a localization observable satisfying the usual Galilei covariance conditions: such a situation cannot be realized within a separable Hilbert space $\mathcal{H}_\mathcal{A}$.

Next, the classical nature of the pointer observable $A_\mathcal{A}$ cannot be reconciled with the fact that the Hamiltonian H of $\mathcal{S} + \mathcal{A}$ represents an observable. To escape this difficulty, one may either exploit the possibility that (in general) the classical nature of $A_\mathcal{A}$ is too strong a condition, or one has to consider a modified description of dynamics in quantum mechanics. Concerning the first option, recall that the classical nature of $A_\mathcal{A}$ was only shown for the special case of measurements of type $\mathcal{M}_U^m$; yet another possible remedy would be to be content with some kind of approximatively classical pointer observables allowing only for approximate objectification. The second option also leaves open two possibilities. Either one maintains the unitary dynamical group but gives up the usual physical interpretation of the Hamiltonian. Or one admits nonunitary evolutions, which may make it impossible to derive the necessity of classical pointer observables, or which may even make them unnecessary.

There are limitations on measurability which are of a more general nature. In particular, the fundamental symmetries of physics give rise to conservation laws (thus restricting the Hamiltonians occurring in nature), which may restrict the set of actually measurable observables.

Finally, the possibilities of preparing physical systems seem to depend on the existence of repeatable measurements used as selection procedures, or filters. Thus an understanding of the preparation of physical systems by means of filters presupposes that both pointer and value objectification do indeed take place in quantum mechanics. As far as continuous observables are involved, there is the additional problem that they do not admit any (unitary) repeatable measurements, which makes it difficult to understand their operational definition.

It may be possible that all the problems mentioned do at least allow for a solution in an approximative sense within quantum mechanics. This option has not been worked out in full; it would have to be based on the notion of POV observables and possibly on a modified conception of dynamics. Some steps in this direction have been done in recent years and will be briefly discussed in Chapter IV of this work, where we shall try to survey various approaches for solving the measurement problem.

Chapter IV

Objectification and Interpretations
of Quantum Mechanics

Survey - The problem of objectification

The heart of the measurement problem in quantum mechanics is the objectification requirement and its implications, which were studied in Chapter III. The reactions to the challenge to solve the objectification problem can be systematically classified according to the following schematic summary, which concentrates on the relevance of the classical nature of the pointer observable. First, in order to achieve the objectivity of the pointer observable after the measurement interaction on the basis of a unitary dynamics, it is necessary that the pointer observable has some classical properties so that superpositions of pointer states corresponding to different pointer positions are excluded . This implication is based on the assumption that the system S is a proper quantum system. Here we introduce the following notations:

$\mathcal{PR}$ probability reproducibility condition
$\mathcal{O}$ objectification requirement
$\mathcal{U}$ unitary measurement coupling
$\mathcal{HO}$ the measurement coupling is generated by an observable
$\mathcal{QS}$ the object system is a proper quantum system
$\mathcal{CP}$ the pointer observable has some classical properties.

The first part of the measurement problem can then be formulated as follows (Theorem III.5.2.1):

(MP1) $\qquad (\mathcal{PR})$ & $(\mathcal{O})$ & $(\mathcal{U})$ & $(\mathcal{QS})$ $\implies$ $(\mathcal{CP})$

The second part of the measuring problem consists of an inconsistency between the probability reproducibility condition, the classical nature of the pointer observable and the assumption that the unitary measurement coupling is generated by an observable of the system $S+A$. Hence we have the following implication (Theorem III.6.2.1)

$$(MP2) \qquad (\mathcal{U}) \ \& \ (\mathcal{HO}) \ \& \ (\mathcal{CP}) \implies \neg(\mathcal{PR}).$$

Implications (MP1) and (MP2) summarize the logical situation encountered with the measurement problem. They show that the very condition $(\mathcal{CP})$ which is necessary for the realization of the measurement $(\mathcal{O})$ precludes the realization of the condition $(\mathcal{PR})$ which defines the concept of measurement. In other words the notion of measurement excludes its very realizability. The clash between the phenomenon of nonobjectivity and the need for a classical description of the apparatus was clearly envisaged by the pioneers of quantum mechanics. They tried to deal with it by emphasizing the methodological necessity of placing a "cut" between the object to be described by the theory and the means of observation. Some representative statements are collected in Section 1 in order to illustrate some variants of the Copenhagen interpretations.

Another reaction to the situation described by (MP1) and (MP2) might have been the claim that quantum mechanics is incomplete and cannot, in particular, provide an exhaustive account of the measurement process. Accordingly one would have denied the phenomenon of nonobjectivity and searched for a deterministic hidden variable description, thus eliminating the measurement problem. In this view quantum mechanics would be merely a statistical ensemble theory (Section 2).

Next, taking seriously quantum mechanics as a complete and even universally valid theory, one has to conclude that objectification does not occur. Indeed, considering the apparatus as a proper quantum system, too, there is no room left for classical properties, so that (MP1) must be read as entailing the negation of the objectification requirement $(\mathcal{O})$. The task of any interpretation of quantum mechanics then would be to explain why objectification *appears* to take place. This is the aim of the many-worlds and similar interpretations (Section 3).

Finally, there are several attempts to maintain the goal of objectification, either *within* quantum mechanics (Section 4), or by going *beyond* this theory (Section 5). Various ideas have been developed to approach objectification, either strictly, or in some approximate sense, within the general structure of ordinary quantum mechanics; the general aim is to achieve – at least *effective* – superselection rules, be it by ad hoc stipulation, or via modified dynamics, environment-induced superselection, or in the sense of unsharp objectification. If classical properties of the apparatus system are taken for granted (thus accepting (MP1)), then one is facing the strange conclusion of (MP2) that the Hamiltonian generator of the unitary measurement dynamics cannot be an observable of $\mathcal{S} + \mathcal{A}$. This consideration shows that a modification of the conception of dynamics is mandatory in this sort of approach.

In approaches going beyond quantum mechanics the hope has been given up that quantum mechanics, in its familiar codification based on a separable Hilbert space, can be maintained as a universally valid theory. Rather, it is argued, one has to take into account systems with infinitely many degrees of freedom and the possibility of continuous superselection rules. This would ultimately lead to theories for macroscopic phenomena, which are significantly different form ordinary quantum mechanics.

The options mentioned will be reviewed, without any claim at completeness. In fact, we are not aware of any thorough resolution of the measurement problem and therefore feel free just to comment briefly on the merits of the various approaches and the questions they still seem to be facing.

1. Historical prelude – Copenhagen interpretations

In speaking of the Copenhagen interpretation of quantum mechanics one is usually referring to the interpretation of quantum mechanics which resulted from the discussions between Niels Bohr, Werner Heisenberg, and Wolfgang Pauli with contributions from Max Born and Johann von Neumann. In very broad terms one may say that these discussions were the first attempts to solve the interpretational problems of quantum mechanics considered as a fundamental theory of individual atomic objects. The classical papers by Born (1926), Heisenberg (1927) and Bohr (1928) are important landmarks in this early development, which led to the systematic treatises by von Neumann (1932) and Pauli

(1933) on the mathematical and conceptual foundations of quantum mechanics.

The acceptance of the probability interpretation for the Schrödinger wave function and the acknowledgement of some fundamental limitations on the applicability of the concepts of classical physics in the description of atomic phenomena were important elements in the Copenhagen interpretation, which, however, never developed into a coherent systematic interpretation of quantum mechanics. Still Bohr's Como lecture may be considered to some extent as a codification of the Copenhagen viewpoint. In fact, as is now well known, the views of Bohr, Heisenberg and Pauli on the interpretation of quantum mechanics were divergent and diffuse to the extent that no unique interpretation could have been built on them and that almost any of the present day interpretations of quantum mechanics can be argued as being a systematic development of some of the "Copenhagen views".

The problem of measurement was already well known to the Copenhagen school. Thus to evaluate the present attempts to solve the measurement problem in their proper historical context, we find it useful to review briefly some of the representative views of the Copenhagen school on the measurement problem. We shall start by presenting the intuitive ideas developed by Bohr and Heisenberg and then go on with the more systematic studies by von Neumann and Pauli. Finally, we recall the related views of London and Bauer and of Wigner.

Bohr. According to Bohr the key to the understanding of the quantum theory was the viewpoint of complementarity which he advocated and developed in a series of articles published during the years 1927 – 1962. Rather than attempting to give a systematic exposition of Bohr's views and to argue which of the *isms* Bohr favoured we provide a few citations from the writings of Bohr which, we think, are revealing with respect to Bohr's account of the measurement process.

> *Bohr 1928*
> The quantum theory is characterized by the acknowledgment
> of a fundamental limitation in the classical physical ideas when
> applied to atomic phenomena. The situation thus created is of
> a peculiar nature, since our interpretation of the experimental

material rests essentially upon the classical physical concepts.
$\cdots$ Quantum postulate implies a renunciation as regards the
causal space-time co-ordination of atomic processes. [But] the
idea of observation belongs to the causal space-time way of
description.
$\cdots$ the complementary character of the description of atomic
phenomena $\cdots$ appears as an inevitable consequence of the
contrast between the quantum postulate and the distinction
between object and agency of measurement, inherent in our
very idea of observation.

Bohr 1939

A clarification of the situation as regards the observation problem in quantum theory $\cdots$ was first achieved after the estab-
lishment of a rational quantum mechanical formalism.
$\cdots$ In the first place, we must recognize that a measurement
can mean nothing else than the unambiguous comparison of
some property of the object under investigation with a corresponding property of another system, serving as a measuring instrument, and for which this property is directly deter-
minable according to its definition in everyday language or in
the terminology of classical physics.

Bohr 1948

An adequate tool for the complementary mode of description
is offered by the quantum-mechanical formalism.

The idea of complementarity of observables first conceived by Bohr
is expressed in the present approach to quantum measurements by the
mutual exclusiveness of instruments according to Sections III.2.1 and
III.7. The classical description of measuring devices is reproduced here
by the requirement of the pointer objectification and of the classical
properties of the pointer observable (Section III.5.2).

Heisenberg. The breakthrough for the proper understanding of the
formalism of quantum mechanics lay, according to Heisenberg (1927),
in the discovery of the uncertainty relations. These relations showed, in
a quantitative way, the limitations on the applicability of classical con-

cepts to the description of atomic phenomena. As mentioned at the end of Section III.7.2, the uncertainty relations at the same time may be interpreted as a relaxation of the complementarity. Though Heisenberg's starting point was more formal than Bohr's, Heisenberg later acknowledged Bohr's viewpoint of complementarity as the fundamental basis of quantum mechanics. With respect to an analysis of the measuring process this appears in the fact that, like Bohr, Heisenberg emphasized the methodological necessity for clearly distinguishing the object under investigation and the applied measuring apparatus. Whereas the object was to be described in terms of quantum mechanics, the measuring apparatus had to be described in classical terms. The following extracts from Heisenberg's writings illustrate these remarks.

Heisenberg 1930

It must also be emphasized that the statistical character of the relation [between the values of two quantum mechanical observables – authors' note] depends on the fact, that the influence of the measuring devices on the system to be measured, is treated in another way than the mutual influence of the parts of the system. ⋯ If one were to treat the measuring device as a part of the system ⋯ then the changes of the states of the system considered above as indeterminate would become determinate. But no use could be made of this determinateness unless our observation of the measuring device were free of undeterminateness. For these observations, however, the same considerations are valid, and we should be forced, for example, to include our own eyes as part of the system, and so on. Finally, the whole chain of cause and effect could be quantitatively verified only if the whole universe were incorporated into the system – but then physics has vanished and only a mathematical scheme remains. *The partition of the world into observing and observed system prevents a sharp formulation of the law of cause and effect.* [authors' emphasis] (The observing system need not always be a human being; it may also be an apparatus, such as a photographic plate, etc.)

Heisenberg 1958

In natural science we are not interested in the universe as a whole, including ourselves, but we direct our attention to some part of the universe and make that the object of our studies. ⋯ it is important that a large part of the universe, including ourselves, does not belong to the object.

⋯ [Before or at least] at the moment of observation our object has to be in contact with the other part of the world, namely the experimental arrangement [which is to be described in terms of classical physics].

⋯ After this interaction has taken place, the probability function contains the objective element of tendency and the subjective element of incomplete knowledge, even if it has been a "pure" case before ⋯ The observation itself changes the probability function discontinuously; it selects of all possible events the actual one that has taken place ⋯ .

⋯ the transition from the "possible" to the "actual" ⋯ takes place as soon as the interaction of the object with the measuring device, and thereby with the rest of the world, has come into play; it is not connected with the act of registration of the result by the mind of the observer. The discontinuous change in the probability function, however, takes place with the act of registration, because it is the discontinuous change of our knowledge in the instant of registration that has its image in the discontinuous change of the probability function.

⋯ quantum theory corresponds to the ideal of objective description of the world as far as possible. Certainly quantum theory does not contain genuine subjective features, it does not introduce the mind of the physicist as a part of the atomic event. But it starts with the division of the world into the "object" and the rest of the world, and from the fact that at least for the rest of the world we use the classical concepts in our description.

The partition of the world into an observed system S and a measuring apparatus A which is inevitable for an objective description of the physical system is reformulated in the present report by the reduction of the state of the compound system $S+A$ into the reduced states of S and A, respectively (Section II.1.2). Within the framework of our approach the apparatus A should be described by means of classical concepts, since only in this way can the objectification of the measuring results be achieved (Section III.5).

von Neumann and Pauli. von Neumann did not accept Bohr's view, shared by Heisenberg, of the necessity of classical language in the description of atomic phenomena. On the contrary, according to von Neumann, quantum mechanics is a universally valid theory which applies equally well to the description of macroscopic measuring devices as to microscopic atomic objects. It may well be that a detailed description of a measuring apparatus is highly complicated but, according to von Neumann, no limitation in principle is known for a description of a measuring apparatus and thus of the whole measuring process in quantum mechanics. As is well known, and already noted in Chapter III, von Neumann developed the quantum mechanical theory of the measuring process in a way that still meets today's standards of rigour. Within his approach von Neumann clearly faced the problem that the object system and the measuring apparatus had to be separated after the premeasurement and that this problem could not be solved within quantum mechanics. To solve the dilemma, von Neumann introduced what is known as the projection postulate, and he argued that the final termination of any measuring process is in the conscious observer, in his becoming aware of the measurement result. Von Neumann also argued that – apart from the fact that in the course of the development of physical theories the borderline between a physical object, like the measuring apparatus, and the conscious observer is as if shifted in the direction of the latter – the conscious observer cannot be included in the domain of any physical theory. We let von Neumann speak (in the 1955 translation), using, however, the technical notations of the present text.

von Neumann 1932

We therefore have two fundamentally different types of interventions which can occur in a system $S \cdots$. First, the arbitrary changes by measurements

(1) $$T \mapsto \mathcal{I}_U(\Re)T.$$

Second, the automatic changes which occur with the passage of time

(2) $$T \mapsto U_t^* T U_t.$$

$\cdots$ In the measurement we cannot observe the system S by itself, but must rather investigate the system $S + A$, in order to obtain (numerically) its interaction with the measuring apparatus A. The theory of measurement is a statement concerning $S + A$, and should describe how the state of S is related to certain properties of the state of A (namely, the positions of a certain pointer, since the observer reads these).
$\cdots$ the measurement or the related process of the subjective perception is a new entity relative to the physical environment and is not reducible to the latter.
$\cdots$ But in any case, $\cdots$, at some time we must say: and this is perceived by the observer.
$\cdots$ Now quantum mechanics describes the events which occur in the observed portions of the world, so long as they do not interact with the observing portion, with the aid of process 2, but as soon as such an interaction occurs, i.e., a measurement, it requires the application of process 1. The dual form is therefore justified.

Another member of the Copenhagen school who developed measurement theory in a systematic way was Wolfgang Pauli. His analysis of the measurement problem is almost literally the same as the one given by von Neumann. Indeed Pauli writes as follows (we here give the 1980 translation):

Pauli 1933

The measurement $\cdots$ generates in general $\cdots$ out of a pure case $\cdots$ a mixture $\cdots$ [cf. $T \mapsto \mathcal{I}_U(\Re)T$]. This result $\cdots$ is of decisive importance for the consistent interpretation of the concept of measurement in quantum mechanics. For this result shows that we arrive at consistent results concerning the system, whatever be the way in which the division between the system to be observed (which is described by wave functions) and the measuring apparatus is made. (Cf. J.v. Neumann (1932), where in Chapter VI this question is discussed in detail.)

$\cdots$ It is possible to express the fact that a definite measuring apparatus will be used in the mathematical formalism of quantum mechanics directly. On the contrary, this is not possible with the stipulation that the measurement should give a definite result $\cdots$ Any statement about a physical fact made with the help of a measuring device (observer or the registration apparatus) which is not counted as part of the system cannot (from the standpoint of mathematical formalism which describes directly only probabilities) present a particular, scientifically not pre-determined act which is to be taken into account by a reduction of the wave-packets $\cdots$ [cf. $T \mapsto \mathcal{I}_U(\Re)T \mapsto N_X^{-2}\mathcal{I}_U(X)T$]. We need not be surprised at the necessity for such a special procedure if we realize that during each measurement an interaction with the measuring apparatus ensues which is in many respects intrinsically uncontrollable.

The theory presented in Chapter III can be seen as a further elaboration of the treatment by von Neumann and Pauli, with the important exception that here the measurement problem is not "solved" with a reference to a conscious observer.

Some elaborations. The analysis of the measuring process as given by von Neumann remained more or less unknown for a long time, perhaps due to its then advanced mathematical presentation. Some authors, like London and Bauer (1939), appreciated this work of von Neumann but they felt the need for a "concise and simple" treatment of the problem.

London and Bauer developed the theory of the quantum mechanical measurement process just in accordance with von Neumann. However, they introduced one significant change into von Neumann's description, namely, by including the conscious observer as a part of the quantum mechanical description of the measurement process. Indeed they considered quantum mechanically a system consisting of the object system S, the measuring apparatus A, and the observer $\mathcal{O}$. To solve the objectification problem, London and Bauer went on to assume that the observer can, by "introspection" and with his "immanent knowledge", always rightly create his own objectivity, and thus identify his own pure state. "I am in the state $P[\rho_k]$" so that, due to correlations, the measuring apparatus A is in the state $P[\Phi_k]$ and thus the object system S is in the state $P[\gamma_k]$.

The London-Bauer interpretation of the measurement according to which the objectification of the measuring result is provided only by the activities of the observer's consciousness was criticized by Wigner in 1961. A philosophical analysis of the London-Bauer approach in the light of Wigner's consideration was given by Shimony (1963). With his "friend paradox" Wigner gave an illustration that after the measurement interaction between S, A and $\mathcal{O}$ the observer $\mathcal{O}$ (Wigner's friend) will always be in a state which indicates a definite measuring result. On the other hand, if the total compound system $S + A + \mathcal{O}$ were treated by means of ordinary quantum mechanics in Hilbert space, then the observer $\mathcal{O}$ would be in a superposition of possible final states, that is, in a state of "suspended animation". For Wigner (1963) this result appears absurd and he concludes that the quantum mechanical equations of motion cannot be linear, if macroscopic objects like A and $\mathcal{O}$ are described. However, no explicit proposal for a nonlinear Schrödinger equation was made. Systematic investigations in this direction started only around 1976 (cf. Section 4.2)

Later Wigner changed his mind and adopted the point of view considered by Zeh (1970) according to which the interaction of the macroscopic system (A or $\mathcal{O}$) with the environment destroys the nondiagonal elements in the density matrix of this system (cf. Section 4.3). Wigner proposed a modified von Neumann-Liouville equation which describes the (exponential) decrease in time of the nondiagonal elements in the density matrix of macroscopic systems and thus should lead automatically to the desired objectification of the measuring result. The consciousness of the observer, which in the London-Bauer approach was assumed to solve this problem is no longer needed for the objectification. Moreover, according to Wigner (1983), the consciousness of the human observer is beyond the scope of quantum physics and classical physics.

2. Ensemble and hidden variable interpretations

The rise of the so-called ensemble and hidden variable interpretations of quantum mechanics is much due to the critique of Albert Einstein and Erwin Schrödinger on quantum mechanics as a fundamental theory of individual atomic objects. The probabilistic nature of quantum mechanics and the incompleteness argument of Einstein, Podolsky, and Rosen (1935) led, on the one hand, to a consideration of quantum mechanics only as a statistical theory of atomic objects, and, on the other hand, to the development of proper completions of the theory. The critique of the projection postulate, as put forward by Margenau (1963), was a further impulse for developing the ensemble interpretations of quantum mechanics. Though much of this critique was later found to be unjustified, these interpretations still have their advocates. In this section we shall briefly review the main ideas of these interpretations.

The statistical interpretations of quantum mechanics can be divided into two groups, the *measurement statistics* and the *statistical ensemble* interpretations (Section III.2.4). These interpretations rely explicitly on the relative frequency interpretation of probability, and in them the meaning of probability is often wrongly identified with the common method of testing probability assertions.

In the measurement statistics interpretation the quantum mechanical probability distributions, like E_T^A, are considered only epistemically as the distributions for measurement outcomes. The concept of state is taken to characterize conceptual infinite sequences Γ_T^A of measurement outcomes $a_1, a_2, \cdots$ such that $E_T^A(X) = \lim_{n \to \infty} \frac{1}{n} \sum_{i=1}^{n} c_X(a_i)$. In this pragmatic view quantum mechanics is only a theory of measurement outcomes providing convenient means for calculating the possible distributions of such outcomes. It may well be that such an interpretation is sufficient for some practical purposes, but it is outside the interest of this treatise to go into any further details, for example, to study the presuppositions of such a minimal interpretation. The measurement problem is simply excluded in such an interpretation.

The statistical nature and the alleged incompleteness of quantum mechanics are apparent in the ensemble interpretation of quantum mechanics. In this interpretation a state T characterizes a conceptual infinite collection Γ_T^S of identical, mutually noninteracting systems $S_1, S_2, \cdots$. The probability measures E_T^A defined by a state T describe the distribution of the values of an observable A among the members S_i of the ensemble Γ_T^S. Accordingly, the number $E_T^A(X)$ is the relative abundance of systems S_i in Γ_T^S having the value of A in the set X.

The ensemble interpretation of quantum mechanics describes individual objects only statistically as members of ensembles. Such an interpretation is built on the idea that each observable has a definite value at all times. Thus no measurement problem occurs in this interpretation. Some merits of the ensemble interpretation of quantum mechanics are put forward, for example, in Refs. 7, 8 and 52. But these merits seem to consist only of a more or less trivial avoiding of the conceptual problems, like the measurement problem, arising in a realistic approach.

The basic idea of the *hidden variable* interpretations of quantum mechanics was to "complete" the probabilistic state description of quantum mechanics with some "hidden variables" to obtain a dispersion-free state description providing the quantum mechanical probabilities as statistical averages over the "hidden variables". Much work has gone into this project, and a number of no-go-theorems emerged restricting the

possible forms of such interpretations. It follows from these theorems
that only the so-called contextual non-local forms of the hidden variable
interpretations are tenable at all. The models of Bohm are the best
known examples in this direction [19,20,21]. They have been further
developed in the stochastic mechanics approach [149]. Another recent
hidden variable formulation of quantum mechanics follows closely the
original ideas of Schrödinger on the matter wave interpretation of the
ψ-function [10]. To reemphasize, these approaches are free from the
measurement problem simply because they do not regard quantum mechanics as a complete description of physical reality. Yet, it may be too
early to estimate whether the hidden variable models can ultimately be
developed into a serious alternative to quantum mechanics.

3. Interpretations without objectification

There are at least three types of interpretations of quantum mechanics which take the measurement problem (MP1) as implying that
no objectification takes place at all in a measurement process. They are
the *many-worlds interpretation*, the *witnessing interpretation* and a special *modal interpretation*. These interpretations share a common formal
feature. They are all concerned with unitary premeasurements $\mathcal{M}_U^m$ of
discrete ordinary observables A, for which the final state $U(\varphi \otimes \Phi)$ of the
compound system $\mathcal{S} + \mathcal{A}$ assumes a biorthogonal decomposition with
respect to the final component states γ_i and Φ_i of the object system
$\mathcal{S}$ and the apparatus $\mathcal{A}$ for any initial (vector) state φ of $\mathcal{S}$ (Section
III.2.3):

$$(1) \qquad U(\varphi \otimes \Phi) = \sum_{ij} \langle \varphi_{ij} | \varphi \rangle \psi_{ij} \otimes \Phi_i = \sum N_i \gamma_i \otimes \Phi_i$$

From the results of Section III.3.2 we know that this is the case exactly
when the premeasurement $\mathcal{M}_U^m$ produces strong correlations between
the component states γ_i and Φ_i. The three interpretations consider
in different ways the state $U(\varphi \otimes \Phi)$ in its biorthogonal decomposition

(1)and they ascribe different roles to the reduced states

(2)
$$T_U \doteq \mathcal{R}_S(P[U(\varphi \otimes \Phi)]) = \sum_i N_i^2 P[\gamma_i]$$

(3)
$$T_{A,U} \doteq \mathcal{R}_A(P[U(\varphi \otimes \Phi)]) = \sum_i N_i^2 P[\Phi_i]$$

Furthermore, in these interpretations the strong correlation premeasurement is taken to be the whole measuring process.

3.1 Many-worlds interpretation.

The first attempt to interpret the quantum mechanical formalism without additional assumptions concerning the objectification was made by Everett (1957) and Wheeler (1957). Everett investigated unitary premeasurements of ordinary discrete observables A and studied the decomposition (1) of the state $\psi' = U(\varphi \otimes \Phi)$ of the compound system $S+A$ into a sum of products of two states γ_i and Φ_i, one referring to the object system S and one referring to the apparatus A, the "observer". The "measurement" is then nothing else but the correlation between the respective state γ_i of the system and the "relative state" Φ_i of the observer who is aware of the system's state γ_i. The large variety of alternatives which coexist in the state ψ' has later been interpreted by some authors as an ensemble of "really" existing "worlds" – an idea which has given rise to the name many-worlds interpretation [56,57,209].

According to Everett's analysis the state (1) which provides strong state-correlations between S and A, can be considered already as a description of the complete measuring process. Any definite measurement outcome is described by a product state $\psi_i' = \gamma_i \otimes \Phi_i$, where γ_i is the state of the system and Φ_i represents the observer (that is, the apparatus and some registration devices) as aware that the system S is in the state γ_i. Hence the wave function $\psi' = \sum N_i \psi_i'$ describes the complete variety of possible final states of a measurement.

Assume that the A-measurement $\mathcal{M}_U^m$ is even a strong value-correlation measurement, so that the vectors γ_i are eigenvectors of A: $A\gamma_i = a_i\gamma_i$ for any $i = 1, \cdots, N$ and for all φ. Then the minimal interpretation and the probability reproducibility condition require that the coefficients $N_i^2 = \langle \varphi | E^A(\{a_i\})\varphi \rangle$ which appear in the decomposition (1) are the probabilities for outcomes a_i in the *final* states of S and A. This postulate was justified in the sense of the relative frequency interpretation of probabilities (Section III.2.4). However, in the present case such an interpretation cannot be given. The numbers N_i^2 cannot be interpreted here as relative frequencies of outcomes a_i in the two Gemenge $\{(N_i^2, P[\gamma_i]) : i = 1, \cdots, N\}$ and $\{(N_i^2, P[\Phi_i]) : i = 1, \cdots, N\}$ which correspond to the decompositions (2) and (3) of the reduced states T_U and $T_{A,U}$, simply because these mixed states play no role in the description of a measurement in the many-worlds interpretation. The whole system $\mathcal{S}+\mathcal{A}$ evolves according to the unitary coupling $\varphi\otimes\Phi \mapsto U(\varphi\otimes\Phi)$ and no state reduction or objectification will take place.

The real positive numbers $p_\varphi(a_i) = N_i^2$ are also in the present situation probabilities in the formal sense. But now the results of Section III.2.4 cannot be used for the justification of a relative frequency interpretation. However, even in this "interpretation without objectification" the relative frequency interpretation of the probabilities $p_\varphi(a_i)$ can be justified in the following sense. Let $\mathcal{S}^{(n)} = S_1 + \cdots + S_n$ be a compound system of n identical, equally prepared systems with states φ and denote the compound state by $\varphi^{(n)}$. If on each system the observable A is measured, that is, the observable $A^{(n)}$ is measured on $\mathcal{S}^{(n)}$, the observer will register a sequence $l = (l_1, \ldots, l_n)$ of n index numbers l_k indicating the values a_{l_k} and store it in his memory. The relative frequency of index numbers k in a "memory sequence" l will be denoted by $f^{(n)}(i, l)$. If one defines a relative frequency operator $F_i^{(n)}$ as in Section III.2.4, it is obvious that $\varphi^{(n)}$ is not an eigenstate of A. However, in the limit of large n one obtains the following result due to Hartle (1968).

THEOREM 3.1.1. *Let $\{S_1, \cdots, S_n\}$ be a set of n equally prepared identical systems S_k with states φ and let $\varphi^{(n)}$ be the state of the compound system $\mathcal{S}^{(n)}$. If $F_i^{(n)}$ is the relative frequency operator for the value a_i appearing in the eigenvalue $(a_{l_1}, \cdots, a_{l_n})$ of $A^{(n)}$, then*

$$\lim_{n\to\infty} \langle \varphi^{(n)}|(F_i^{(n)} - p_\varphi(a_i))^2 \varphi^{(n)}\rangle = 0$$

for any $i = 1, \cdots N$ and for all φ.

It has been pointed out by Ochs (1977) that the strong and somewhat unrealistic premises of Hartle's proof (pure state φ, equally prepared and independent systems) can be considerably relaxed and re-placed by more realistic assumptions. Furthermore Ochs emphasized that Theorem 1 is essentially a "law of large numbers" which shows that quantum probabilities fulfil this important requirement. Theorem 1 means that for a given observer, that is, an apparatus plus a registration device, the relative frequency $f^{(n)}(i,l)$ of values a_i will approach (for large n) the probability $p_\varphi(a_i)$ in *almost* every memory sequence $l = (l_1, \ldots, l_n)$ [57]. At first glance this result is somewhat surprising since none of the n systems S_i with preparation φ possesses an A-value a_k in an objective sense. However, this argument does not invalidate the relevance of the mentioned result, since within the present interpretation for the measuring process the objectification of the measured observable is not required in any sense.

In addition to these more fundamental problems we mention some further consequences of the present interpretation illustrating the advantages and disadvantages of the many-worlds interpretation. For details we refer to the literature [57].

For the description of the measuring process the consciousness of the observer is not needed. Automata for the registration of memory sequences are completely sufficient. Several automata can be used for the formation of a measurement chain without thereby changing the registered results in the respective memory sequences. The results of the automata persist throughout the whole chain. Moreover two automata are also allowed to communicate with each other since this exchange of information will never lead to a paradoxical situation like the paradox of Wigner's friend.

Quantum theory in Hilbert space in its present form does not describe the reality which we usually have in mind but some reality which is composed of many distinct worlds. The observer is only in one of these worlds. For this reason he is unable to recognize the full determinism of the totality of many worlds which are described by ψ', but only one part of it which is governed by a probability law, the justification of which is given by Theorem 1.

It follows from these arguments that quantum mechanics can be applied to an individual system. Every automaton, that is, a measuring apparatus with a memory sequence, describes the system correctly according to the probability laws. For this reason quantum theory can be applied even to the universe, its creation and its development. The state of the universe corresponds to a decision tree with an enormous number of branches. Every measuring process splits the world again into many alternative components, but the whole description is completely consistent.

The many-worlds interpretation seems to be a proper interpretation of quantum mechanics with its usual axioms. It does not presuppose the pointer objectification and the value objectification, hence avoiding the problem (MP1). On the other hand, it makes use of Theorem 1 as a legitimation of the probability interpretation, since this theorem need not to be understood here in the sense of an ignorance interpretation. However, the price for this interpretation of quantum measurement theory is a very strange ontology: quantum mechanics does not describe any longer the *one* world in which we are living but at the same time the totality of all possible worlds.

The merits of the many-worlds interpretation consist in the observation that the quantum mechanical formalism provides an interpretation of the theory, including the relative frequency interpretation of formal probabilities, if the strong correlations between the object state and the memory state of the observer are considered as already representing the "measurement". However, the additional statement that the possible "worlds" really exist, seems to be rather irrelevant for the following two reasons: first the "many-worlds hypothesis" is not compatible with any physically acceptable concept of reality (cf. for example, Ref. 54); secondly the existence of the alternative worlds cannot be falsified by any quantum mechanical experiment. Finally the many-worlds interpretation as presented above relies on the assumption that the biorthogonal decomposition (1) is unique. This is the case only if all the initial probabilities N_i^2 are different from each other. The consistency of the many-worlds interpretation cannot be claimed before it is shown that this assumption can be avoided.

3.2 Witnessing interpretation.

Another interpretation of quantum mechanics which goes beyond the minimal interpretation and which aims to avoid the objectification problem is the witnessing interpretation of quantum mechanics proposed by Kochen (1985).

The main ideas of the witnessing interpretation can be expressed directly in the context of the measurement theory where it applies the biorthogonal decomposition (1) of the measurement interaction U in an A-measurement $\mathcal{M}_U^m$. Indeed in the witnessing interpretation only such unitary mappings U for which $\sum_i N_i \gamma_i \otimes \Phi_i$ is the biorthogonal decomposition of $U(\varphi \otimes \Phi)$ are considered as possible measurement interactions for A. We assume from now on that $\mathcal{M}_U^m$ is such. In that case the biorthogonal decomposition (1) of $U(\varphi \otimes \Phi)$ and the canonical decompositions (2) and (3) of T_U and $T_{A,U}$ are the natural decompositions with respect to the measurement process. The core of this interpretation is given by the following assumption.

(W) *When the biorthogonal decomposition of $U(\varphi \otimes \Phi)$ assigns the mixed states $\sum N_i^2 P[\gamma_i]$ and $\sum N_i^2 P[\Phi_i]$ to the two interacting systems S and A, then S actually has exactly one of the properties $P[\gamma_i]$ as witnessed by A, and A has the corresponding property $P[\Phi_i]$ as witnessed by S. In other words, if $U(\varphi \otimes \Phi) = \sum N_i \gamma_i \otimes \Phi_i$, then S and A are in one of the corresponding states γ_i and Φ_i.*

The above assumption (W), cited from Ref. 118, shows that the very heart of the witnessing interpretation is in the claim that when the compound system $S + A$ is in the vector state $U(\varphi \otimes \Phi) = \sum N_i \gamma_i \otimes \Phi_i$, then S and A are in one of the pure states $P[\gamma_i]$ and $P[\Phi_i]$, $i = 1, \cdots, N$. As is clear from the analysis of Section III.5, such an assumption does not follow from the biorthogonal decomposition, that is, from the assumption that $\mathcal{M}_U^m$ is a strong state-correlation measurement. The assumption (W) is indeed an extra assumption going beyond the minimal interpretation. Moreover, it is exposed to the arguments presented in Section II.2.5 which show that the correlated mixed states T_U and $T_{A,U}$ do not admit an ignorance interpretation.

The witnessing interpretation of quantum mechanics shows some similarities with the London-Bauer interpretation of quantum mechanics as described in Section 2. In the London-Bauer approach the role of a witness is explicitly given to an observer who possesses a sufficient mental capacity. However, the witnessing interpretation was introduced with the aim "of showing that there is a consistent view of the formalism as describing an objective world of individual interacting systems" [118]. Hence the identification of the two interpretations seems unjustified.

The conceptual foundations of the witnessing interpretation has not yet been worked out in a systematic way. Hence it is beyond the scope of the present review to study the consistency of this interpretation. As in the case of the many-worlds interpretation, the uniqueness of the biorthogonal decomposition (1) is also crucial for the witnessing procedure. It remains to be seen whether this assumption can be removed. By the results of Section III.5 the assumption (W) has the same function as the assumption that the pointer observable is classical. Thus the question remains whether the witnessing interpretation also faces a problem similar to (MP2).

Measurement theoretical foundations of the witnessing interpretation have been studied in Ref. 125. Görnitz and von Weizsäcker (1987 a,b) have tried to present a continuation of the Copenhagen interpretation which – in their view – considers quantum theory as a description of human knowledge. They understand Kochen's perspective interpretation as a similar continuation of the Copenhagen view, which in several interesting points coincides with their own approach. We note finally that a "resolution of the measurement problem" on the basis of ideas similar to the witnessing interpretation has been developed by Dieks (1988, 1989, 1990).

3.3 A modal interpretation.

A further interpretation of quantum mechanics which goes beyond the minimal interpretation but which avoids the objectification problem is a modal interpretation of quantum mechanics developed by van Fraassen (1981, 1990). This interpretation considers quantum mechanics as "*a pure theory of the possible*, with testable, empirical implications for what actually happens" [202].

Consider a physical system $\mathcal{S}$, and let E and T be any of its observables and states. The modal interpretation distinguishes between two types of propositions: *the value-attributing propositions* – an observable E has value X, to be denoted (E, X) – and *the state-attributing propositions* – a measurement of E leads to a result in X, to be denoted $[E, X]$. In accordance with the minimal interpretation a state T *makes* the state-attributing proposition $[E, X]$ *true* if $E_T(X) = 1$. The heart of the interpretation is to characterize the truth of a value-attributing proposition, that is, to answer the question which value-attributions (E, X) are true in a given state T. The truth of (E, X) will not be identified with the truth of $[E, X]$. To explain this the following definition is needed: a state T' is *possible relative to T* if and only if $tr[T'P] = 1$ whenever $tr[TP] = 1$ for any projection operator P. A vector state $P[\varphi]$ is thus possible relative to T exactly when φ is in the range of T. In particular, if $P[\varphi]$ occurs in a decomposition of T, then $P[\varphi]$ is possible relative to T (cf. Section II.2.4). The modal interpretation then starts with the following postulate.

(P) *Given that system $\mathcal{S}$ is in a state T, there is a certain pure state $P[\varphi]$ which is possible relative to T, and such that for all observables E pertaining to $\mathcal{S}$:*
a) a state-attribution $[E, X]$ is true if and only if T makes it true;
b) a value attribution (E, X) is true if and only if $P[\varphi]$ makes $[E, X]$ true.

The item *a)* in (P) is a special case of the minimal interpretation, whereas *b)* is an additional assumption specific to the modal interpretation. It should be emphasized that the pure state $P[\varphi]$ appearing in *b)* does not represent an extra dynamic state of $\mathcal{S}$. It plays only a bookkeeping role in delimiting the value-attributions (E, X) which are true about $\mathcal{S}$ in its (dynamic) state T. If $\mathcal{S}$ is in a pure state $T = P[\varphi]$, then the true value-attributions are given exactly by the true state-attributions, and the state T can be identified with a maximal collection of true value-attributions (E, X). In general such an identification is, however, not possible. The bookkeeping devices are called the *value states*, and they make the value-attributions true. In order to distinguish clearly the value states from states, states shall be called *dynamic states* in this section. Dynamic states make the state-attributions true.

Let $\mathcal{M}_U^m$ be a strong state-correlation measurement of A. The dynamic states of $\mathcal{S}+\mathcal{A}$, $\mathcal{S}$ and $\mathcal{A}$ after the measurement are then given by Equations (1), (2) and (3), provided that φ is the initial dynamic state of $\mathcal{S}$. These states determine which state-attributions are true, and they delimit the value-attributions which could be true now. In particular, the pointer eigenstates $P[\Phi_i]$ are possible relative to $T_{\mathcal{A},U}$ whenever $N_i^2 \neq 0$. The value state $P[\Phi_k]$ makes the value-proposition $(A_{\mathcal{A}}, a_k)$ true, and, due to the strong state-correlation, the probability for (A, a_k) being true in the corresponding value state γ_k is $\langle \gamma_k \mid E^A(\{a_k\})\gamma_k\rangle$. If the measurement would be also a strong value-correlation measurement, then this probability were equal to one, that is, the proposition (A, a_k) would also be true. However, one cannot specify what the value states of $\mathcal{A}$ and $\mathcal{S}$ are. The possible value states relative to $T_{\mathcal{A},U}$ can, of course, be determined, and the measurement scheme provides the probability distribution $i \mapsto N_i^2$ for some of those possibilities, namely for the pointer eigenstates. Faced with this situation the modal interpretation adds the following assumption concerning measurements [202].

(M) *If* $U(\varphi \otimes \Phi) = \sum N_i \gamma_i \otimes \Phi_i$ *is the dynamic state of* $\mathcal{S}+\mathcal{A}$ *after the A-measurement* $\mathcal{M}_U^m$, *then* N_i^2 *is the probability that the value states of* $\mathcal{S}$ *and* $\mathcal{A}$ *are* γ_i *and* Φ_i, *respectively.*

This assumption is consistent with the postulate (P), and it leads to the following result when applied to the strong value-correlation measurements of A.

(R) *If* $U(\varphi \otimes \Phi) = \sum N_i \gamma_i \otimes \Phi_i$ *is the dynamic state of* $\mathcal{S}+\mathcal{A}$ *after the A-measurement* $\mathcal{M}_U^m$, *then the probability that the value-attribution* (A, a_k) *is true, given that the value-attribution* $(A_{\mathcal{M}}, a_k)$ *is true, equals one.*

The measurement assumption (M) of the modal interpretation leads to the result (R), which is *as if the objectification had taken place*. The assumption (M) is crucial here. It claims that for any initial state φ

of $\mathcal{S}$ the only possible pure states relative to T_U and $T_{A,U}$ are $P[\gamma_i]$ and $P[\Phi_i]$, respectively. The probability for other value states which are possible relative to T_U and $T_{A,U}$ is assumed to be zero. If $\mathcal{S}$ has no (nontrivial) classical properties, then assumption (M) seems to play the same role as that of the classical pointer observable A_A of $\mathcal{M}_U^m$.

One may argue that the modal interpretation of quantum mechanics does not need the measurement assumption (M). Indeed result (R) does not presuppose (M). However, without this assumption the concept of measurement seems to remain rather vague. As in the case of the witnessing interpretation it is too early to judge the consistency of the modal interpretation, and, in particular, its ability to circumvent the measurement problem (MP2). This interpretation applies the minimal interpretation in a particularly striking way. It is used to select in the measurement context the decomposition $\sum N_i^2 P[\Phi_i]$ as the relevant decomposition of the reduced apparatus state $T_{A,U}$. The modal interpretation seems to introduce also a strange feature for the theory of compound systems, as applied in the measurement theory. Indeed in the final dynamic state $T_{A,U}$ of the apparatus the value-attributions (A_A, a_k) are possibly true, but in the final dynamic state $U(\varphi \otimes \Phi)$ of the compound system no corresponding value-attribution $(I \otimes A_A, a_k)$ can be true.

Concluding this section we note that there exist some other proposed interpretations (without objectification) of quantum mechanics which try to maintain some kind of value assignment without adopting a hidden variable point of view. Examples are the consistent histories approach of Griffiths (1985), the proposal of Omnès (1988a,b,c) or the decoherent histories theory of Gell-Mann and Hartle (1990). These attempts are partly motivated by the idea that the quantum mechanical probabilities refer to both prediction and retrodiction [2]. It may be noted that there are severe limitations on the objectifiability of past events, which make any realistic interpretation of retrodiction highly problematic [177]. The difficulties of reconciling such interpretations with the requirements of a realistic interpretation of quantum mechanics and with the rules of ordinary logic are pointed out by d'Espagnat (1989, 1990). In fact, an "objectification" of nonobjective properties in the sense of a (hypothetical, mental) value assignment is possible only if the laws of classical logic are weakened in the sense of quantum logic [144,145] (cf. Section II.2.3).

4. Aiming at objectification within quantum mechanics

Instead of sacrificing either the completeness of quantum mechanics or the goal of objectification, one may consider possibilities of approaching objectification by means of explaining in some way or other the effective classical nature of the pointer observable. In view of the obstacles (MP1) and (MP2), this requires at least some modification of the usual axiomatics, but may perhaps allow one to remain *within* the framework of quantum mechanics. Attempts in this spirit are the subject of the present section. It should be recalled that in the present treatise quantum mechanics is understood as being based on the structure of a separable complex Hilbert space. Hence the introduction of superselection rules and the consideration of a modification of the dynamical axioms are regarded as taking place *within* quantum mechanics. In this sense there are more radical approaches going *beyond* quantum mechanics, which leads to a questioning of the universal validity of this theory (Section 5). Some authors are more liberal in the use of the term quantum mechanics, including, for instance, its C^*-algebraic generalizations.

4.1 Ad hoc superselection rules.

Sometimes it has been argued that the object system S itself, as well as the apparatus system A, have naturally restricted sets of observables so that not all self-adjoint operators correspond to observables. Such limitations on measurability may be due to fundamental conservation laws [6,212], or due to the limited number of actually existing interactions [174] (cf. Section III.7.3). While in this way one circumvents the implication (MP1) by giving up (QS), it is an open question whether objectification can be achieved in this way. Such a solution would not have the status of a theoretically well-founded approach either but would rest upon accidental facts as long as there were no theory of the fundamental interactions. This, admittedly vague, option is mentioned here just for completeness as one logical possibility of dealing with (MP1).

A more effective attempt at dealing with (MP1) is the (more or less ad hoc) consideration of superselection rules. That is, one assumes that the pointer, being an observable of a *macroscopic* system, should be a classical observable [15,106,107,109,206]. Then the objectification ($\mathcal{O}$) is ensured since the pointer observable is objective throughout. However, the consistency problem (MP2) still persists and requires a modification of the dynamics in the sense of at least giving up ($\mathcal{HO}$) [15,206].

While superselection rules can easily be incorporated into the Hilbert space framework of quantum mechanics, they nevertheless represent an element which does not follow from the standard axiomatics. Thus the problem remains of providing a convincing theoretical explanation for such restrictions on the sets of observables and states. Furthermore, the consistency problem (MP2) may be interpreted as an indication of the fact that the irreversibility inherent in measurement has not been taken into account properly. In fact, the unitary evolution on a (separable) Hilbert space is quasiperiodic [164] and generally time inversion invariant, while irreversibility requires a breaking of this fundamental time inversion symmetry. Accordingly there are many model considerations aiming at an explanation of the effectively irreversible evolution of the system $\mathcal{S} + \mathcal{A}$ into a state equivalent to that required by objectification. Usually the equivalence of states is stipulated with respect to a restricted class of macroscopic observables of $\mathcal{A}$. The effective irreversibility is achieved by taking account of the macroscopic nature and ergodic properties of $\mathcal{A}$ [44,47,92,134,181,207]. In this way one obtains an approximate description of the dynamics in terms of a Markovian master equation, which would be sufficient as far as the macroscopic observables (and their functions) are concerned. Moreover, these approaches allow one to investigate the thermodynamic limit, thus affording a bridge to theories dealing with infinite systems (Section 5). A more recent attempt along these lines considers irreversibility via dynamical instability properties on the level of the von

Neumann–Liouville equation from which an effective restriction of the set of states can be derived; in this way the two important features of the measurement problem – irreversibility and classical properties – are shown to be interrelated [143].

4.2 Modified dynamics.

It may be possible to avoid the implication $(\mathcal{CP})$ of (MP1) by means of a modification of the dynamical axiom of quantum mechanics (hence trying the negation of $(\mathcal{U})$) in such a way as to achieve *dynamically induced superselection rules*. One may assume a spontaneous stochastic process to supersede the ordinary unitary evolution. The corresponding generalization of the Schrödinger equation is interpreted as representing the autonomous dynamics of isolated systems.

The need for giving up the linearity of dynamics in order to reach objectification was recognized as late as 1963 by Wigner [214] but no explicit proposal for a nonlinear Schrödinger equation was made at this time. Systematic investigations in this direction started around 1976 [18,62,70,78,163].

The presently best known and perhaps most elaborated example of this type of approach is the *unified dynamics* theory of Ghirardi, Rimini and Weber [17,76,77]. In this theory the von Neumann–Liouville equation for a quantum mechanical n-particle system is modified by adding a linear term which models a *spontaneous localization* process taking place at random times:

$$(1) \qquad i\hbar \frac{d}{dt} T \;=\; [H,T]_- \;+\; \sum_{k=1}^{n} \lambda_k \left(\int_{\Re} A_{q_k} T A_{q_k}^+ dq \;-\; T \right).$$

The integral term is known in the context of the theory of unsharp measurements as the nonselective operation representing the reduced state of the object system after an unsharp position measurement. The operators A_{q_k} are of the form $A_{q_k} = \sqrt{\alpha/\pi}\, exp\left[-\alpha(Q_k - q_k)^2\right]$, where Q_k represents (a component of) the position operator of the k-th particle.

Equation (1) represents a quantum dynamical semigroup with a non-self-adjoint generator, so that the irreversibility needed for measurement is built in from the outset. The mean rates (λ_k) of the localization processes can be chosen small enough so that systems having only a few number of constituents practically evolve according to the Schrödinger equation, while the localization events become noticeable only for systems with a macroscopically large number of constituents. In this way the localization of a macroscopic system is an observable which is *dynamically objectified* practically at any instant of time. Consequently in the measurement context system $\mathcal{A}$ is a proper quantum system itself, without any need of an ad hoc introduction of classical pointer observables; on the contrary, postulate $(\mathcal{CP})$ is no longer a necessity, since objectification $(\mathcal{O})$ is ensured due to a dynamical restriction of the set of states of $\mathcal{A}$ (that is, by giving up $(\mathcal{U})$).

It should be kept in mind that the spontaneous localization process corresponds to an *unsharp* position observable, so that the induced Gemenge is a continuous family of nonorthogonal and only unsharply localized states. It is an open question whether on the basis of generalized stochastic dynamics one encounters a consistency problem of the form (MP2). While this theory provides an interesting step towards a unified quantum mechanical description of microscopic and macroscopic systems, it would still be desirable to establish a theoretical basis for its new dynamical principle. In particular, the question has been raised of whether the theory can be formulated so as to make it Lorentz covariant [13,14]. Recent work provides steps towards its incorporation into relativistic quantum theory [75].

A systematic and more general investigation of stochastic modifications of the unitary dynamics has been carried out in the works of Gisin and Pearle [79,163]. In particular, it was shown that Equation (1) can be equivalently written as a stochastic nonlinear Schrödinger equation. It should be noted that the recourse to nonlinearity in quantum mechanics bears with it somewhat dangerous consequences such as the possible existence of superluminal signals [80,162]. However, it seems possible to formulate such theories in accordance with special relativity and causality [79]. Apart from this remark, the same modest reservations as those raised with respect to the Ghirardi-Rimini-Weber theory apply also to the more general stochastic dynamical theories. Finally we note that the feasibility of experimental tests of dynamical objectification and of nonlinear dynamics in general seems to be an open and difficult problem [161,190,208].

4.3 Environment approaches.

It has been argued that due to its macroscopic nature a measuring apparatus $\mathcal{A}$ must be considered as an open system. According to Zeh, its interaction with the environment cannot be neglected but rather may be made responsible for the emergence of classical properties and irreversibility as required for measurement. Incorporating the environment $\mathcal{E}$ into the description and reducing the unitary $(\mathcal{U})$, Hamiltonian $(\mathcal{HO})$ dynamics of $\mathcal{S} + \mathcal{A} + \mathcal{E}$ to that of $\mathcal{S} + \mathcal{A}$ leads to a nonunitary evolution of the latter system.

As pointed out by Zurek (1981, 1982, 1984), the interaction of $\mathcal{A}$ with the environment should be responsible for specifying the pointer observable (the "pointer basis") and effecting the transition to the $\mathcal{S} + \mathcal{A}$-mixture required for objectification (Equation (10) of Section III.5.2). By means of idealized models, Zurek was able to show that a quantum nondemolition type of interaction between $\mathcal{A}$ and $\mathcal{E}$ leads to a kind of monitoring of the pointer through the environment. As a consequence, the reduced state of $\mathcal{S} + \mathcal{A}$ is indeed approximately (or quasi-) diagonal in the pointer basis. The last statement holds to a good degree of accuracy and with large Poincaré recurrence times if the environment possesses a large number of constituents. One may say that the coherence originally present in the state of $\mathcal{S} + \mathcal{A}$ after the measurement interaction is not destroyed but *dislocalized* into the many degrees of freedom of the environment. The fact that the apparatus $\mathcal{A}$ is always found in a mixture of pointer states can be interpreted as the appearance of an *environment-induced* superselection rule. In the past few years more realistic models in this spirit have been devised, showing the emergence of classical properties in macroscopic systems due to interactions with their environment [81,91,113,117,205]. These models exploit the dissipative or amplifying influences of $\mathcal{E}$ on the (apparatus) system $\mathcal{A}$ under consideration. In the measurement context, the reduced $\mathcal{S} + \mathcal{A}$ dynamics assumes the form of a quantum dynamical semigroup similar to that of the Ghirardi-Rimini-Weber theory (cf. also Ref. 215).

It must be emphasized that these approaches do not reach the goal of objectification $(\mathcal{O})$ since the *effective* environment-induced superselection rules are obtained by voluntarily neglecting the – *practically* unobservable – degrees of freedom of the environment. Indeed reading (MP1) and (MP2) as statements about the total system $\mathcal{S} + \mathcal{A} + \mathcal{E}$, these obstacles remain perfectly valid. In particular, the resulting $\mathcal{S} + \mathcal{A}$-mixture does not admit an ignorance interpretation with respect to the pointer basis. Ignoring the environment then amounts only to an *apparent* abandoning of the postulate $(\mathcal{U})$ of unitary dynamics. Hence what can be achieved could at best be called *quasi-objectification*.

The advocates of the environment approach are well aware of the limitations just mentioned. For instance, Joos and Zeh (1985), as well as Zurek (1982), emphasize very explicitly that the quantum mechanical coherence is not destroyed but only displaced. This process of displacement can be described very clearly in information theoretical terms [159]. To conclude, we suggest that the environment approaches may be taken to provide important insights into the framework of interpretations without objectification (Section 3). In particular, the explanation of a preferred pointer basis seems to offer a way towards an understanding of the robustness of macroscopic properties. Furthermore, insofar as the models studied allow for taking thermodynamic-type limits, they may also serve as prototypical examples for any approach towards the measurement problem, especially for those considering infinite systems and thus going beyond quantum mechanics. In fact an infinite environment will lead to infinite recurrence times and strict superselection rules (cf. Section 5).

4.4 Unsharp objectification.

The *unsharp objectification* programme is not intended, in the first instance, to provide a new, independent approach to the measurement problem. It rather emerges as a reflection on important developments which seem to provide the proper tools for a conceptually sound description of a quasi-classical domain within quantum mechanics.

It was often argued that classical properties of quantum systems should emerge if these systems are macroscopic in some sense. If this view turned out to be true, then it would render the ad hoc recourse to superselection rules (Section 4.1) an admissible idealization. Yet the usual textbook technique of taking the limit $\hbar \rightarrow 0$ is far too crude a procedure. It is merely a mathematical operation without any operational justification. Certainly the "smallness" of $\hbar$ *in some sense* is important, but its precise meaning should follow from physical requirements. An operational criterion should result from a characterization of "classical properties", "classical behaviour", etc. Again, it seems difficult to provide generally acceptable definitions of these terms. One may require (approximately) deterministic trajectories, either within certain subspaces of "classical" states, or for expectation values of "macroscopic" observables. Examples of the first option are classical descriptions based on (generalized) coherent states; for a brief survey and precise definitions, cf. Ref. 170. The second method is employed, for instance, by Ludwig (1987) or by Daneri, Loinger and Prosperi (1962); cf. also Ref. 175. Instead of reviewing the various attempts, we present a number of examples illustrating, in a self-explanatory way, that quantum mechanics does possess a quasi-classical level of description. The essential new point in these examples is the decisive role ascribed to *unsharp* observables which are represented by positive operator valued measures. An important quasi-classical feature of unsharp quantum observables is the existence of *coexistent* sets of noncommuting observables. This coexistence can be achieved by means of introducing a sufficiently large degree of unsharpness. The resulting blurring of potential interference can be utilized in the explanation of an approximately classical behaviour of macroscopic observables.

As a first example we mention the possibility of joint unsharp measurements of quantum mechanical position and momentum observables [4,49,102,176,182]. Such quantum mechanical phase space measurements behave like classical trajectory determinations provided the involved position and momentum unsharpnesses are macroscopically large [30]. Hence classical trajectories can be approximately realized in terms of unsharp quantum observables [115,198,199].

Next, it may be recalled that Bell's inequalities are satisfied *within* quantum mechanics if coexistent triples of unsharp observables are considered [40,117]. Thus, in the Einstein–Podolsky–Rosen experiment one may assume, without running into contradictions, the simultaneous *unsharp* objectivity of noncommuting spin observables.

Finally, a "surprising quantum effect" discovered recently [1] displays a classical rotation behaviour of quantum mechanical spin observables in the course of a sequence of weak measurements. This classical feature has been shown to be a consequence of an Einstein–Podolsky–Rosen type correlation between the involved spin N system and the weak measurement device [32]. Due to the huge unsharpness of the weak measurement the entanglement between object and apparatus is not completely destroyed but, on the contrary, guarantees the appearance of the surprising quantum effect: the occurrence of apparently forbidden readings lying far beyond the spectrum of the observables measured. The large number of constituents of the spin N system ensures that noncommuting spin components simultaneously have relatively well-defined values. Taking into account that such a weak measurement defines an unsharp observable, one finds a natural explanation of the phenomenon under consideration [32], one lesson being that the emergence of classical features may even be considered as a macroscopic quantum effect.

The measurement unsharpness represented by unsharp observables must not be confused with a mere inaccuracy; on the contrary, it is rooted in a genuine quantum indeterminacy inherent in the measuring device. In the light of the above examples it seems appropriate to understand macroscopic observables as unsharp observables with a macroscopic degree of intrinsic unsharpness [30,33]. The magnitude of this unsharpness induces a kind of *natural coarse-graining* in the sense that it fixes a scale on which the macroscopic observable assumes fairly well-defined values.

In this context another observation is of great importance: the existence of *quasi-classical* states. In a model of a quantum mechanical amplifier Glauber (1986) described a very peculiar type of "macroscopic" oscillator states which have the property of being *quasi-diagonal* with respect to various observables: position, momentum, energy, or phase-space observables. The term quasi-diagonality refers to the exponential decay of nondiagonal elements with increasing separation from diagonal terms. It is plausible to assume that measurements of macroscopically unsharp observables performed on such states do not allow one to detect interference effects proving the nonobjectivity of some of these observables. Hence the above-mentioned noncommuting observables may be assumed to be unsharply objective since this assumption presumably cannot be disproved.

Applying these ideas to the measurement context leads to the following scenario. Presumably any realistic pointer observable is an intrinsically and macroscopically unsharp observable, hence a macroscopic observable in the sense explained above. It may turn out that the interaction of the measuring device with the environment forces the apparatus to remain in some quasi-classical states in which, in particular, the pointer observable is quasi-diagonal. Then it becomes impossible to "see" (in the sense of macroscopic observation) the apparatus in a superposition of macroscopically distinguishable states. The "objectification" reached in this way is an essentially unsharp one and certainly does not require classical properties in the sense of superselection rules.

The description of macroscopic observables as macroscopically unsharp observables is not only plausible but has been shown to be a necessary consistency condition for the embedding of macroscopic deterministic theories into quantum mechanics (cf. Ref. 137, Chapter X.2.5). Some of the models referred to in the preceding subsections seem to fit in well with the spirit of unsharp objectification [81,91,117,121,122,205]. In these models the quasi-classical behaviour of a macroscopic quantum system $\mathcal{A}$ is shown to result from interactions with its environment $\mathcal{E}$. The pointer states are essentially nonorthogonal, though still almost orthogonal. This is related to the fact that they refer to a macroscopically unsharp pointer observable. It appears likely that a closer analysis of these models will show that the resulting $\mathcal{A}$-states are indeed quasi-classical states.

Obviously the unsharp objectification proposal leaves a number of questions to be settled. Some clarifications concerning the operational meaning of unsharpness have been obtained in recent investigations, and these lead to a conception of an *unsharp quantum reality*. The resulting generalized measurement theoretical concepts, such as that of approximate repeatability, allow for an extension of the quantum nondemolition idea to continuous and even unsharp observables [37]. In this way a precise understanding of the nondemolition monitoring of coherent pointer states through some environment can be achieved.

It must be emphasized that unsharp objectification is indeed a relaxation of the objectification requirement. Accordingly this option does not lead, via (MP1), to classical observables of the apparatus in the strict sense, but it is based rather on approximately classical features of macroscopic quantum observables. To conclude, the unsharp objectification programme offers further developments and improvements in the environment approaches, contributing, in particular, the notion of quasi-classical mixtures of pointer states, on the basis of a new conception of macroscopic observables as unsharp observables. The results to be expected may ultimately serve to provide illustrations of the general analysis given by Ludwig (1987) concerning the relationship between macro-theories and many-body quantum mechanics.

5. Approaching objectification beyond quantum mechanics

As mentioned in the preceding section, the incorporation of superselection rules needs to be supplied with some physically convincing motivation. The attempted justifications reviewed so far have only led to *effective* superselection, either by recourse to modified dynamics or via monitoring through the environment. The only known physical mechanism for producing strict superselection rules is by means of spontaneous symmetry breaking, which may occur in systems having infinitely many degrees of freedom. With regard to the measurement problem we discuss separately two classes of approaches which consider the apparatus as an infinite system: first we review attempts working with continuous superselection rules *within* the Hilbert space framework, after which we turn to those approaches which start out with more general structures such as C^*-algebras or others.

5.1 Continuous superselection rules.

Approaches based on discrete superselection rules (Section 4.1) are somewhat artificial as they do not allow for continuous trajectories of the classical (pointer) observables involved. In fact, as soon as one starts formulating concrete models, one quite inevitably introduces some continuous observables such as position, which are declared classical observables. A very elaborate theory is that by Sherry and Sudarshan (1978, 1979) in which the measurement process is described as an interaction between the quantum mechanical object system S and the classical measuring device A. The latter system is given a quantum mechanical description by means of embedding its observables into the set of self-adjoint operators of some Hilbert space. The classical nature of A is preserved by stipulating that its trajectory variable corresponds to a (continuous) classical (hence superselection) observable. In this theory the dynamical problem (MP2) is taken into account and leads to a requirement of classical integrity of the localization variable. This approach does not aim at justifying the classical nature of the measuring device, but it provides a very instructive example of how to reconcile classical and quantum mechanical descriptions in the context of measurement.

The "many–Hilbert spaces" theory by Machida and Namiki (1980) is an attempt to explain the emergence of classical properties of measuring devices due to their macroscopic nature. It is argued that the energy and the number of constituents of an apparatus A, being a macroscopic and therefore open system, are not well-defined. Hence the state of A should be a mixture of states with different definite particle numbers, these numbers being distributed in a relatively narrow range around the macroscopically large mean number n_0. In the limit $n_0 \to \infty$ one obtains a state operator which can be represented as an average with respect to a continuous size parameter. In this way one effectively performs a transition to a direct integral of Hilbert spaces in which macroscopic observables are defined as continuous averages of microscopic observables. As pointed out by Araki (1980), this procedure leads to macroscopic observables inducing continuous superselection rules. Measurement models involving such continuous superselection rules have been subsequently developed further by other authors (see, for instance, Ref. 71).

The price to be paid for the implementation of continuous super-selection rules in realistic models is that one is generally dealing with nonseparable Hilbert spaces [5,90,168]. Furthermore the dynamical consistency problem (MP2) still persists. The fact that in the case of continuous superselection rules the Hamiltonian generator of the dynamical group cannot be an observable was noticed already by Piron (1976).

5.2 Algebraic attempts.

From a formal point of view, the continuous-superselection approaches are based on the choice of a particular representation of some C^*-algebra of observables. Indeed our distinction between approaches within or beyond "quantum mechanics" corresponds to the distinction of algebras of observables constituting factors of type I, II or III respectively.

The C^*-algebra approach was developed primarily as a rigorous way of dealing with systems having infinitely many degrees of freedom, as they occur in relativistic quantum field theories or in (quantum) statistical physics. It allows for an extension of quantum mechanics with the same set of axioms as the latter. Due to this structural identity the extended theory has itself been called (algebraic) quantum mechanics. For a concise presentation of this theory, as well as a survey of the relevant literature we may mention the monograph [169].

The possibility of solving the objectification problem (MP1) by means of classical properties of the apparatus A rests on the description of A as a system having a quasi-local C^*-algebra of observables. Such an algebra has infinitely many inequivalent (Gelfand-Neumark-Segal) representations. (Instructive and "elementary" illustrations of this phenomenon are given in Refs. 29 and 185.) Related to this is the existence of *disjoint* states on these representations, which serve as candidates for a pointer basis [101]. The interaction between A and some quantum mechanical object system S is represented by means of a one-parameter group of automorphisms acting on the tensor product of the respective W^*-algebras. The measurement models elaborated by

Hepp (1972) demonstrate that, by a suitable choice of the interaction, probability reproducibility as well as repeatability can be achieved via strong value-correlation. Furthermore, due to the disjointness of the pointer states, the dynamics lead, in the infinite-time limit, to a state which is equivalent (in the sense described in Section III.5.2) to the mixture required by the objectification condition. Again, the disjointness of the pointer states guarantees that the ignorance interpretation can be applied to this mixture. The macroscopic observable associated with the disjoint pointer states is given as the space-average of a microscopic local observable, thus in a similar way as in the model of Machida and Namiki (cf. Section 5.1). A systematic formulation of the measurement conditions in the C^*-algebraic framework, together with some elaborations of Hepp's models are given in Ref. 211.

It must be noted, however, that the time evolution, being time-inversion invariant, can lead to objectification only in the infinite-time limit. Indeed for any finite time one can find an observable of $\mathcal{A}$ which shows significant interference terms proving the persisting coherence of the $\mathcal{S}+\mathcal{A}$ state [12]. This fact demonstrates that the algebraic approach towards measurement has its own dynamical problem, reminiscent of (MP2). In order to achieve a gradual approach within finite time towards the objectification, one therefore has to break the time-inversion symmetry of the evolution. Lockhart and Misra (1986) propose to incorporate the required irreversible behaviour of $\mathcal{A}$ by means of a suitable selection of the von Neumann algebra of observables of $\mathcal{A}$ in such a way that a certain causality requirement is satisfied. The theory developed by these authors thus interconnects the phenomena of objectification and irreversibility, showing that an understanding of the latter provides also an explanation of the former in the framework of algebraic quantum mechanics.

In recent years Primas (1990b) has strongly advocated the approach sketched out and elaborated a programme towards an *individual* description of the measurement process. Defining a quantum *object* as a system with no Einstein–Podolsky–Rosen correlations with its environment, it follows that a proper quantum system is an object exactly when its environment is a classical system [178]. (This result has a structure similar to that of our (MP1), the difference being that we do not start with such a restrictive definition of an object.) According to Primas, the measurement problem then consists of finding a suitable representation of the algebra of observables of $S + A$ such that S "lives" in a classical environment. In this picture the reduced dynamics of S turns out to be a stochastic process, described by a nonlinear stochastic Schrödinger equation (cf. Section 4.2). Due to enormous technical difficulties, the development of models along these lines is at the moment in a rather preliminary stage [172,217].

Algebraic quantum mechanics entails powerful mathematical tools for formulating generalized quantum theories. In particular, it seems to provide a framework wide enough to allow for a consistent and thorough theory of the measurement process. Yet the proposals presented so far make use of assumptions (concerning, for instance, the mechanisms utilized for breaking the time-inversion symmetry, cf. Ref. 133), which may be regarded as debatable. However, taking literally the actual infinity (of the number of constituents, or of degrees of freedom) in order to solve foundational problems of a physical theory seems hardly acceptable in itself. In our view, the consideration of infinite limits in a physical theory is an idealization made in a state where there is a lack of a more detailed knowledge about the situation under investigation; it is an admissible idealization as long as no inconsistency or conflict with observation arises. Whatever position one adopts with respect to this question, it should be kept in mind that taking such limits is a well-exercised technique for approaching a simplified description of a physical problem, yielding hints for a more comprehensive solution. In this sense the measurement models of algebraic quantum mechanics are also valuable contributions to approaches *within* (ordinary) quantum mechanics, serving as guides for a better understanding of the nature of macroscopic observables and their – possibly only approximate – classical behaviour.

5.3 Operational approaches.

Instead of using formal extensions of ordinary quantum mechanics some authors have elaborated generalizations of quantum mechanics on the basis of an operational reconstruction of the theory. First, the quantum logic approach is open, from the outset, to the possible existence of superselection rules; it also allows one to formulate continuous superselection rules [15,90,168]. This approach shows that there are no a priori operational reasons inherent in the general language structure of quantum mechanics, which would either exclude or require the existence of superselection rules. This is an illustration of the fact that the measurement problem is, indeed, specific to the irreducible case of ordinary Hilbert space quantum mechanics.

Another type of operational approach starts with an analysis of the general statistical structure of physical theories [135]. This procedure allows one to investigate, in very general terms, the relationship between objective, deterministic theories for macroscopic phenomena and quantum mechanics as a theory for microscopic systems. If it can be shown that the former cannot be derived from a many-body extrapolation of the latter, then the universal validity of quantum mechanics is lost, and there is no reason to expect that measuring devices belong to the domain of quantum mechanics [136].

Contrary to the usual textbook wisdom it has been argued as a result of careful analysis that macrophysical theories, like thermodynamics or classical mechanics, cannot be derived as approximate limiting cases of quantum mechanics. Such theories should rather be regarded as theories in their own right. Yet as macroscopic systems are aggregates of microscopic systems, it is important to understand the interrelation between quantum mechanics and macroscopic theories. This programme has been carried out to quite a large extent in the form of a consistent embedding of macrotheories into a quantum mechanical many-body theory [9,137]. The effect of such an embedding can be described, roughly speaking, as restricting the sets of states and observables of the many-body quantum theory so that the remaining structure allows for the emergence of a classical (deterministic) behaviour of the macroscopic observables.

The conclusion to be drawn from this approach is that there is no way of *deriving* a classical behaviour of macroscopic observables from quantum mechanics, but the classical features of macroscopic systems can be consistently *described* in an approximate way within quantum mechanics. In this way the measurement problem has disappeared at the expense of giving up the universal validity of quantum mechanics.

Chapter V

Concluding Discussion

In this treatise we have tried to develop a systematic exposition of the quantum theory of measurement. The operational language of general observables and instruments allows for a concise definition of the notion of measurement, from which it is possible to derive necessary and sufficient conditions for the realizability of quantum mechanical measuring processes.

The quantum theory of measurement is motivated by the idea of the universal validity of quantum mechanics, according to which this theory should be applicable, in particular, to the measuring process. Hence one would expect, and most researchers in the foundations of quantum mechanics have done so, that the problem of measurement should be solvable *within* quantum mechanics. The long history of this problem shows that, in spite of many important partial results, there seems to be no straightforward route towards its solution. This general impression is confirmed in the present work by means of a number of no-go-theorems.

After an attempt to localize the measurement problem within a variety of interpretations of quantum mechanics (Chapter I), we described the phenomenon of nonobjectivity (Chapter II) which raises the question of what the term measurement could possibly mean in quantum mechanics (Survey of Chapter III). The notion of measurement proposed in this work (Section III.1) is based on two fundamental requirements, probability reproducibility and objectification. The former condition is needed in order to specify which processes can be regarded as premeasurements of a given observable. The latter requirement stipulates that a measurement should lead to a definite result, which is problematic in view of the phenomenon of nonobjectivity.

The probability reproducibility condition, as well as other measurement theoretical postulates such as strong correlations, repeatability or ideality, can be formally incorporated into the quantum theory of measurement in a consistent manner (Sections III.2–4). The fulfilment of these postulates in actual measurements will generally meet obstacles due to dynamical and other limitations on measurability (Sections III.6-7). However, the heart of the quantum mechanical measurement problem must be seen in the second fundamental requirement, objectification (Section III.5). In the context of normal unitary premeasurements with a nondegenerate pointer observable – a situation which is most frequently considered in the literature – we arrived at the following dramatic conclusions. Firstly, objectification can be obtained if and only if the pointer is a classical observable (Theorem III.5.2.1). Hence the objectification requirement necessitates the presence of superselection rules for the measuring apparatus. Moreover, the presence of a classical, discrete and nondegenerate pointer observable entails that the respective apparatus cannot be a quantum system as constituted in (Galilei or Poincaré) spacetime. To reconcile the constitution of a measuring apparatus via systems of imprimitivity with the objectification requirement seems to require continuous superselection rules so that separable Hilbert spaces would no longer provide a sufficient basis for quantum mechanics (Section III.5.2).

The situation becomes even more serious if the *simultaneous* fulfilment of the probability reproducibility and objectification requirements is taken into consideration. In fact the conventional conception of measurement dynamics as generated by an observable of the object plus apparatus compound system, together with the assumption of a classical pointer observable, excludes the realization of the probability reproducibility (Theorem III.6.2.1). Hence we are facing the disastrous conclusion that the very concept of a premeasurement, as characterized by the probability reproducibility condition, precludes its realization as a measurement in the sense of objectification (cf. the Survey of Chapter IV). In this way the two basic measurement postulates turn out to contradict each other in the context of what may be called the standard measurement model, but most likely also in more general situations. This result leads to yet another serious consequence: insofar as the preparation of physical systems requires the possibility of performing measurements, it seems impossible to understand the process of preparation within quantum mechanics (Section III.8).

Consequently the quantum theory of measurement seems to be incapable of reaching its goal: while ordinary quantum mechanics does allow for the description of measuring processes, it does not seem to provide an explanation of their realizability. The question now is how to evaluate these remarkable conclusions and their implications.

A systematic and complete account of the possible reactions to and attitudes towards the measurement problem is offered by the logical structure of this problem as summarized in the Survey of Chapter IV. It proves useful as a guide to the various contributions found in the literature, which are the subject of Chapter IV. The most radical option perhaps is to question the very basis of the whole argumentation: in the hidden variable approaches (Section IV.2) the phenomenon of nonobjectivity is not accepted as a feature of physical reality but rather taken as evidence against the completeness of quantum mechanics, with the implication that the objectification problem becomes irrelevant within quantum mechanics. The opposite extreme position is taken in the interpretations without objectification (Section IV.3), which maintain the universal validity of quantum mechanics. These proposals have been seen not to be without problems; at the least they refer to a rather bizarre ontology.

Other more realistic approaches towards an understanding of objectification within the framework of quantum mechanics are those invoking classical observables, either ad hoc, or in the sense of environment–induced superselection rules (Section IV.4). The former proposal is exposed to the no-go-theorems mentioned: realistic measurement models based on classical pointer observables seem to require a framework more general than provided by separable Hilbert spaces. The latter, environment approach does at best lead to effective superselection, valid only as long as the environment is voluntarily ignored. Recent developments in the theory of unsharp observables have made conceivable a new option within quantum mechanics, that of unsharp objectification, which is based on the possible emergence of quasi-classical properties of macroscopic quantum systems (Section IV.4).

More radical approaches try to solve the measurement problem by means of some modifications of the axiomatics of ordinary quantum mechanics. Thus they do not regard this theory as being universally valid. We have reviewed the modified dynamics attempts as one possibility allowing one to maintain the usual Hilbert space framework (Section IV.4.2). The operational approaches (Section IV.5) consist of invoking more general theories, making essential reference to infinite (open) systems; this requires a reformulation of the entire measurement problem.

This list of attempts to approach the objectification problem illustrates the lively development going on in the quantum theory of measurement. In our view it is too early to decide which of the options described will ultimately lead to a thorough picture of quantum mechanical measurements. The final answer will require further investigations, especially in problems concerning macroscopic quantum systems and the (quasi-) classical limit of quantum mechanics.

It seems appropriate at this point to note the striking analogy between the quantum mechanical measurement problem and the problem of irreversibility in statistical mechanics. The latter problem consists of explaining the observed irreversible and stochastic behaviour of macroscopic systems in view of the time-inversion symmetry of the underlying microscopic dynamical equation. Similarly, the measurement process appears to be a stochastic process, leading irreversibly to definite outcomes, while according to quantum mechanics the underlying evolution of the object plus apparatus system is governed by a Schrödinger equation. In accordance with this parallelism, some of the approaches to the measurement problem resemble very much certain techniques belonging to the realm of statistical physics.

The relevance of the macroscopic nature of the measuring apparatus to the measurement problem was envisaged in several instances in the present work. One may show that the difficulties in detecting interferences between macroscopically distinct (pointer) states increase with an increasing number of constituents of the apparatus [114,127]. In analogy with the strategy of ergodic theory it was claimed that quantum chaos, as the source of irreversibility, is necessary for ensuring the consistency of a solution, within quantum mechanics, of the the measurement problem [165,166]. Further connections between irreversibility and objectification are exploited in the approaches to the measurement problem reviewed in Section IV.5. In fact the recourse to macroscopic or infinite systems has much in common with the thermodynamic limit procedures of statistical physics [72,128,133,143,156,171,172,173].

The parallelism between the objectification and the irreversibility problems allows one to take advantage of the experiences gathered from the latter century-old problem, in an estimation of the possibilities for dealing with the former. This immediately seems to suggest caution in considering any ad hoc changes of quantum mechanics. On the other hand, this could also be seen as an invitation to consider both problems in a framework more general than that of ordinary quantum mechanics. Perhaps the most elaborated treatment of the question of macroscopic systems in this general spirit is given by Ludwig (1987) (cf. our Section IV.5.3). It does seem that we have to live in a situation which does not yet allow of any definite conclusions.

With these remarks we feel we have indicated important issues for future research into the physics underlying quantum measurements and at the same time illustrated the philosophical relevance of our theme. A general philosophical discussion of the problem of measurement, together with an evaluation of some of the approaches reviewed in Chapter IV can be found in d'Espagnat's (1987,1989) work which aims at an elucidation of the quantum mechanical conception of reality. We have not touched upon a methodological problem related to the self-referential nature of a quantum theory of measurement [46,167,173]. In fact from a methodological point of view the measuring process does not belong to the domain of quantum mechanics but rather serves to constitute the semantics of this theory. It is the requirement of the semantical completeness of quantum mechanics which stipulates that the very (measuring) processes providing operational definitions of the concepts of the theory must be describable in terms of the theory. This semantical completeness, which is illustrated by the pseudo-realistic figure below, induces a logical situation similar to that encountered with Gödel's theorem [83]. To avoid inconsistencies within a universally valid quantum mechanics, it is argued, the theory cannot be applied to yield

143

a complete description of a measurement situation. Rather one has to accept that part of the process of measurement remains unanalyzed [167]; in other words, according to this point of view one has to distinguish between two levels of description: the endophysical (ontic) and the exophysical (epistemic) levels [172,173], analogously to the distinction between object language and metalanguage in logic. In our opinion these ideas deserve to be taken seriously; but they also require further elaboration towards rigorous formalization before their far-reaching implications can be properly estimated.

With these short philosophical remarks we shall conclude our treatise on the quantum theory of measurement. In the Joensuu-1990 Symposium Jean-Marc Lévy-Leblond has argued that the measurement problem was discussed so many times in the literature that any further "contribution" could only contribute to increasing the general confusion. On the other hand, this very conference, as well as many others, showed in an impressive way the present status of research in this field, displaying a variety of promising lines of development. One may indeed expect further new and important insights into the foundations of quantum mechanics in the near future. We therefore hope that the reader will find our effort useful, at least as providing a language and framework for future investigations into the quantum theory of measurement.

Figure 1. *Illustration of the semantical completeness of quantum mechanics. Adapted from L. S. Penrose, R. Penrose, The British Journal of Psychology 49 (1958) 31.*

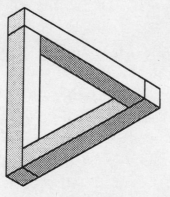

da capo al fine

References

1. AHARONOV, Y., ALBERT, D. Z., CASHER, A., and VAIDMAN, L. (1987). Surprising Quantum Effect. *Physics Letters A 124,,* 199-203. [IV.4.4]

2. AHARONOV, Y., BERGMANN, P. G., and LEBOWITZ, J. L. (1964). Time Symmetry in the Quantum Process of Measurement. *Physical Review 134B*, 1410-1416. [IV.3.3]

3. AHARONOV, Y., and BOHM, D. (1961). Time in the Quantum Theory and the Uncertainty Relations for Time and Energy. *Physical Review 122*, 1649-1658. [III.6]

4. ALI, S. T. (1985). Stochastic Localization, Quantum Mechanics on Phase Space and Quantum Space-Time. *La Rivista del Nuovo Cimento 8(11)*, 1-128. [IV.4]

5. ARAKI, H. (1980). A Remark on Machida-Namiki-Theory of Measurement. *Progress of Theoretical Physics 64*, 719-730. [IV.5.1]

6. ARAKI, H., and YANASE, M. (1960). Measurements of Quantum Mechanical Operators. *Physical Review 120*, 622-626. [III.7]

7. BALLENTINE, L. E. (1970). The Statistical Interpretation of Quantum Mechanics. *Reviews of Modern Physics 42*, 358-381. [IV.2]

8. BALLENTINE, L. E. (1988). What do we learn about quantum mechanics from the theory of measurement? *International Journal of Theoretical Physics 27*, 211-218. [IV.2]

9. BARCHIELLI, A., LANZ, L. and PROSPERI, G. M. (1982). A Model for the Macroscopic Description and Continual Observation in Quantum Mechanics. *Il Nuovo Cimento 72B*, 79-121. [IV.5.3]

10. BARUT, A. O. (1990). Quantum theory of single events. In: Joensuu (1990), pp. 31-46. [IV.2]

11. BELL, J. S. (1966). On the Problem of Hidden Variables in Quantum Mechanics. *Reviews of Modern Physics 38*, 447-452. [I.2]

12. BELL, J. S. (1975). On Wave Packet Reduction in the Coleman-Hepp Model. *Helvetica Physica Acta 48*, 93-98. [IV.5.2]

145

13. BELL, J. S. (1987). Are There Quantum Jumps? In: *Schrödinger. Centenary of a Polymath*, Cambridge University Press, Cambridge. (Reprinted in: J. S. Bell, *Speakable and Unspeakable in Quantum Mechanics*, Cambridge University Press, 1987.) [IV.4.2]

14. BELL, J. S. (1989). Towards An Exact Quantum Mechanics. In: *Themes in Contemporary Physics. Essays in Honor of Julian Schwinger's 70th Birthday*, S. Deser, R. J. Finkelstein, eds., World Scientific, Singapore, pp. 1-26. [IV.4.2]

15. BELTRAMETTI, E., and CASSINELLI, G. (1981). *The Logic of Quantum Mechanics*. Addison-Wesley, Reading, Massachusetts. [I; II; III]

16. BELTRAMETTI, E., CASSINELLI, G., and LAHTI, P. J. (1990). Unitary Measurements of Discrete Quantities in Quantum Mechanics. *Journal of Mathematical Physics 31*, 91-98. [II; III]

17. BENATTI, F., GHIRARDI, G. C., RIMINI, A., and WEBER, T. (1987). Quantum Mechanics with Spontaneous Localization and the Quantum Theory of Measurement. *Il Nuovo Cimento 100B*, 27-41. [IV.4.2]

18. BIAŁYNICKI-BIRULA, I., and MYCIELSKI, J. (1976). Nonlinear Wave Mechanics. *Annals of Physics 100*, 62-93. [IV.4.2]

19. BOHM. D. (1952). A Suggested Interpretation of the Quantum Theory in Terms of "Hidden Variables". Part I: *Physical Review 85*, 166-179; Part II: *Physical Review 85*, 180-193. [IV.2]

20. BOHM, D., and BUB, J. (1966). A Proposed Solution of the Measurement Problem in Quantum Mechanics by a Hidden Variable Theory. *Reviews of Modern Physics 38*, 453-469. [IV.2]

21. BOHM, D., and HILEY, B. J. (1989). Non-Locality and Locality in the Stochastic Interpretation of Quantum Mechanics. *Physics Reports 172, No. 3*, 93-122. [IV.2]

22. BOHM, D., and VIGIER, J. P. (1954). Model of the Causal Interpretation of Quantum Theory in Terms of a Fluid with Irregular Fluctuations. *Physical Review 96*, 208-216. [I]

23. BOHR, N. (1928). The Quantum Postulate and the Recent Development of Atomic Theory. *Nature, 121*, pp. 580-590. [IV.1]

24. BOHR, N. (1939). Causality Problems in Atomic Physics. In: *New Theories in Physics*, Paris, 1939, pp. 11-45. [IV.1]

25. BOHR, N. (1948). On the Notions of Causality and Complementarity. *Dialectica 1*, 312-319. [IV.1]

26. BOHR, N. (1949). Discussion with Einstein on Epistemological Problems in Atomic Physics. In: *Albert Einstein: Philosopher-Scientist*, P. A. Schilpp, ed., Library of Living Philosophers, Evanston, Illinois, pp. 663-688. [I.2]

27. BORN, M. (1926). Zur Quantenmechanik der Stoßvorgänge. *Zeitschrift für Physik 37*, 863-867. [IV.1]

28. BROWN, H. R. (1986). The Insolubility Proof of the Quantum Measurement Problem. *Foundations of Physics 16*, 857-870. [III.5]

29. BUB, J. (1988). How to Solve the Measurement Problem of Quantum Mechanics? *Foundations of Physics 18*, 701-722. [IV.5.2]

30. BUSCH, P. (1982). Unbestimmtheitsrelation und simultane Messungen in der Quantentheorie. Dissertation, Cologne. English translation: Indeterminacy Relations and Simultaneous Measurements in Quantum Theory. *International Journal of Theoretical Physics 24*, 63-92. [I; III.7; IV.4.4]

31. BUSCH, P. (1985). Momentum Conservation Forbids Sharp Localisation. *Journal of Physics A 18*, 3351-3354. [III.7.3]

32. BUSCH, P. (1988). Surprising Features of Unsharp Quantum Measurements. *Physics Letters A 130*, 323-329. [IV.4]

33. BUSCH, P. (1990). Macroscopic Quantum Systems and the Objectification Problem. In: Joensuu (1990), pp. 62-76. [III.5.2; IV.4.4]

34. BUSCH, P., CASSINELLI, G., and LAHTI, P. J. (1990). On the Quantum Theory of Sequential Measurements. *Foundations of Physics 20*, 757-775. [III.2; III.3.6]

35. BUSCH, P., GRABOWSKI, M., and LAHTI, P. J. (1989). Some Remarks on Effects, Operations and Unsharp Measurements. *Foundations of Physics Letters 2*, 331-345. [II.1; III.7.2]

36. BUSCH, P. and LAHTI, P. J. (1989). The Determination of the Past and the Future of a Physical System in Quantum Mechanics. *Foundations of Physics 19*, 633-678. [III.8]

37. BUSCH, P. and LAHTI, P. J. (1990a). Some Remarks on Unsharp Quantum Measurements, Quantum Non-Demolition and All That. *Annalen der Physik 47*, 369-382. [III.7; IV.4.4]

38. BUSCH, P., and LAHTI, P. J. (1990b). Completely Positive Mappings in Quantum Dynamics and Measurement Theory, *Foundations of Physics 20*, 1429-1439. [III.2]

39. BUSCH, P., LAHTI, P. J. and MITTELSTAEDT, P. (1991). Some Important Classes of Quantum Measurements and their Information Gain. *Journal of Mathematical Physics* [III.2.3; 3.6; 3.7]

40. BUSCH, P., and SCHROECK, F. E., JR. (1989). On the Reality of Spin and Helicity. *Foundations of Physics 19*, 807-872. [I; III.7; IV.4.4]

41. CASSINELLI, G., and LAHTI, P. J. (1989). The Measurement Statistics Interpretation of Quantum Mechanics: Possible Values and Possible Measurement Results of Physical Quantities. *Foundations of Physics 19*, 873-890. [III.2.4]

42. CASSINELLI, G., and LAHTI, P. J. (1990). Strong Correlation Measurements in Quantum Mechanics. *Il Nuovo Cimento 105B*, 1223-1233. [III.3.4]

43. CASSINELLI, G., and ZANGHI, N. (1984). Conditional Probabilities in Quantum Mechanics II. *Il Nuovo Cimento 79 B*, 141-154. [III.5.1]

44. CINI, M., DE MARIA, M., MATTIOLI, G., and NICOLO, F. (1979). Wave Packet Reduction in Quantum Mechanics: A Model of a Measuring Apparatus. *Foundations of Physics 9*, 479-500. [IV.4.1]

45. COLOGNE (1984). *Recent Developments in Quantum Logic*. P. Mittelstaedt and E. W. Stachow, eds., Bibliographisches Institut, Mannheim, 1985. [I.2]

46. DALLA CHIARA, M. L. (1977). Logical Self-Reference, Set Theoretical Paradoxes and the Measurement Problem in Quantum Mechanics. *Journal of Philosophical Logic 6*, 331-347. [V]

47. DANERI, A., LOINGER, A., and PROSPERI, G. M. (1962). Quantum Theory of Measurement and Ergodicity Conditions. *Nuclear Physics 33*, 297-319. (Reprinted in [210].) [IV.4.1]

48. DAVIES, E. B. (1970). On the Repeated Measurement of Continuous Observables in Quantum Mechanics. *Journal of Functional Analysis 6*, 318-346. [III.7.1]

49. DAVIES, E. B. (1976). *Quantum Theory of Open Systems*. Academic Press, London. [I; II; III]

50. DAVIES, E. B., and LEWIS, J. T. (1970). An Operational Approach to Quantum Probability. *Communication on Mathematical Physics 17*, 239-259. [III.3.6; III.7]

51. DE BROGLIE, L. (1953). *La Physique Quantique Restera-t-elle Indéterministique?* Gauthier-Villars, Paris. [I]

52. D'ESPAGNAT, B. (1971). *Conceptual Foundations of Quantum Mechanics*. Benjamin, Addison-Wesley, Reading, Massachusetts, 2nd. edition, 1976. [IV.2]

53. D'ESPAGNAT, B. (1987). Empirical Reality, Empirical Causality and the Measurement Problem. Foundations of Physics 17, 507-529. [V]

54. D'ESPAGNAT, B. (1989). Are there Realistically Interpretable Local Theories? *Journal of Statistical Physics 56*, 747-766. [IV.3;V]

55. D'ESPAGNAT, B. (1990). Towards a Separable "Empirical Reality"? *Foundations of Physics 20*, 1147-1172. [IV.3.3]

56. DEWITT, B. S. (1970). Quantum Mechanics and Reality. *Physics Today 23*, 30-35. [IV.3.1]

57. DEWITT, B. S. (1971). The Many-Universes Interpretation of Quantum Mechanics. In: *Foundations of Quantum Mechanics*, B. d'Espagnat, ed., Academic Press Inc., New York. [III.2.4; IV.3.1]

58. DEWITT, B. S., and GRAHAM, N. (1973). *The Many-Worlds Interpretation of Quantum Mechanics*. Princeton University Press, Princeton. [III.2.4; IV.3.1]

59. DIEKS, D. (1988). The Formalism of Quantum Theory: An Objective Description of Reality? *Annalen der Physik 45*, 174-190. [IV.3.2]

60. DIEKS, D. (1989). Quantum Mechanics without the Projection Postulate and its Realistic Interpretation. *Foundations of Physics 19*, 1397-1423. [IV.3.2]

61. DIEKS, D. (1990). Resolution of the Measurement Problem through Decoherence of the Quantum State. *Physics Letters A142*, 439-446. [IV.3.2]

62. DIÓSI, L. (1989). Models for Universal Reduction of Macroscopic Quantum Fluctuations. *Physical Review A 40*, 1165-1174. [IV.4.2]

63. EINSTEIN, A. (1936). Physics and Reality. *Journal of the Franklin Institute 221*, 349-382. [I.2]

64. EINSTEIN, A., PODOLSKY, B., and ROSEN, N. (1935). Can Quantum-Mechanical Description of Physical Reality be Considered Complete? *Physical Review 47*, 777-780. [I.2; III.1; IV.2]

65. EVERETT, H. (1957). The Theory of the Universal Wave Function. In: DeWitt and Graham (1973). [I; III.2.4; IV.3.1]

66. FEINSTEIN (1958). *Foundations of Information Theory.* Mc Graw-Hill, 1955. [III.4.1]

67. FINE, A. (1969). On the General Quantum Theory of Measurement. *Proceedings of the Cambridge Philosophical Society 65*, 111-122. [III.1]

68. FINE, A. (1970). Insolubility of the Quantum Measurement Problem. *Physical Review D2*, 2783-2787. [III.5.2]

69. FOULIS, D., and RANDALL, C. H. (1978). The Operational Approach to Quantum Mechanics. In: *The Logico-Algebraic Approach to Quantum Mechanics III*, C. A. Hooker, ed., Reidel, Dordrecht, pp. 167-201. [I.2]

70. FRENKEL, A. (1990). Spontaneous Localizations of the Wave Function and Classical Behavior. *Foundations of Physics 20*, 159-188. [IV.4.2]

71. FUKUDA, R. (1990). Macrovariables and the Theory of Measurement. In: *Tokyo (1989)*, pp. 124-134. [IV.5.1]

72. GAVEAU, B., and SCHULMAN, L. S. (1990). Model Apparatus for Quantum Measurements. *Journal of Statistical Physics 58*, 1209-1230. [V]

73. GDAŃSK (1987). *Problems in Quantum Physics – Gdańsk 1987.* L. Kostro, A. Posiewnik, J. Pykacz and M. Žukowski, eds., World Scientific, Singapore. [I.2]

150

74. GELL-MANN, M., and HARTLE, J. B. (1990). Quantum Mechanics in the Light of Cosmology. In: *Tokyo (1989)*, pp. 321-343. [IV.3.3]

75. GHIRARDI, G. C., GRASSI, R., and PEARLE, P. (1990). Relativistic Dynamical Reduction Models: General Framework and Examples. In: Joensuu (1990), pp. 109-123. [IV.4.2]

76. GHIRARDI, G. C. and RIMINI, A. (1990). Old and New Ideas in the Theory of Quantum Measurement. In: *Sixty-Two Years of Uncertainty. Historical, Philosophical and Physics Inquiries into the Foundations of Quantum Physics.* A. I. Miller, ed., Plenum Press, New York. [IV.4.2]

77. GHIRARDI, G. C., RIMINI, A., and WEBER, T. (1986). Unified Dynamics for Microscopic and Macroscopic Systems. *Physical Review D 34*, 470-491. [III.6; IV.4.2]

78. GISIN, N. (1984). Quantum Measurement and Stochastic Processes. *Physical Review Letters 52*, 1657-1660. [III.7; IV.4.2]

79. GISIN, N. (1989). Stochastic Quantum Dynamics and Relativity. *Helvetica Physica Acta 62*, 363-371. [IV.4.2]

80. GISIN, N. (1990). Weinberg's Non-linear Quantum Mechanics and Supraluminal Communications. *Physics Letters A 143*, 1-2. [IV.4.2]

81. GLAUBER, R. J. (1986). Amplifiers, Attenuaters and the Quantum Theory of Measurement. In: *Frontiers of Quantum Optics.* E. R. Pike and S. Sarkar), eds., Adam Hilger, Bristol. [IV.4.4]

82. GLEASON, A. M. (1957). Measures on the Closed Subspaces of a Hilbert Space. *Journal of Mathematics and Mechanics 6*, 885-893. [I.2, II.2.2]

83. GÖDEL, K. (1931). Über formal unentscheidbare Sätze der Principia Mathematica und verwandter Systeme I. *Monatshefte für Mathematik und Physik 38*, 173-198. [V]

84. GÖRNITZ, T., and VON WEIZSÄCKER, C. F. (1987a). Remarks on S. Kochen's Interpretation of Quantum Mechanics. In: Joensuu (1987), pp. 357-368. [IV.3.2]

85. GÖRNITZ, T., and VON WEIZSÄCKER, C. F. (1987b). Quantum Interpretations. *International Journal of Theoretical Physics 26*, 921-937. [IV.3.2]

86. GRABOWSKI, M. (1990). Quantum Measurement and Dynamics. *Annalen der Physik 47* 391-400. [III.6]

87. GRAHAM, N. (1973). The Measurement of Relative Frequency. In: DeWitt and Graham (1973), pp. 229-253. [III.2.4]

88. GRIFFITHS, R. B. (1984). Consistent Histories and the Interpretation of Quantum Mechanics. *Journal of Statistical Physics 36*, 219-272. [IV.3.3]

89. GROENEWOLD, H. J. (1971). A Problem of Information Gain by Quantum Measurements. *International Journal of Theoretical Physics 4*, 327-338. [III.4]

90. GUENIN, M. (1966). Axiomatic Foundations of Quantum Theories. *Journal of Mathematical Physics 7*, 271-282. [IV.5.1]

91. HAAKE, F., and WALLS, D. F. (1987). Overdamped and Amplifying Meters in the Quantum Theory of Measurement. *Physical Review A 36*, 730-739. [IV.4.3]

92. HAAKE F., and WEIDLICH, W. (1968). A Model for the Measuring Process in Quantum Theory. *Zeitschrift für Physik 213*, 451-465. [IV.4.3]

93. HADJISAVVAS, N. (1981). Properties of Mixtures of Non-Orthogonal States. *Letters on Mathematical Physics 5*, 327-332. [II.2.4]

94. HALMOS, P. (1988). *Measure Theory*. Springer Verlag, New York. [III.4.2]

95. HARTLE, J. B. (1968). Quantum Mechanics of Individual Systems. *American Journal of Physics 36*, 704-712. [III.2.4; IV.3.1]

96. HEISENBERG, W. (1927). Über den anschaulichen Inhalt der Quantentheoretischen Kinematik and Mechanik. *Zeitschrift für Physik 43*, 172-198. [I; III.7.2; IV.1]

97. HEISENBERG, W. (1930). *Die physikalischen Prinzipien der Quantentheorie*. S. Hirzel Verlag, Leipzig. [IV.1]

98. HEISENBERG, W. (1958). *Physics and Philosophy*. Harper & Row, New York. [IV.1]

99. HELLWIG, K. E. (1971). Measuring Processes and Additive Conservation Laws. In: *Foundations of Quantum Mechanics*, B. d'Espagnat, ed., Academic Press, New York, pp. 338-345. [III.7.3]

100. HELSTROM, C. W. (1976). *Quantum Detection and Estimation Theory*. Academic Press, New York. [III.4]

101. HEPP, K. (1972). Quantum Theory of Measurement and Macroscopic Observables. *Helvetica Physica Acta 45*, 237-248. [IV.5.2]

102. HOLEVO, A. S. (1982). *Probabilistic and Statistical Aspects of Quantum Theory*. North Holland Publishing Corporation, Amsterdam. [I.2]

103. INGARDEN, R. (1976). Quantum Information Theory. *Reports of Mathematical Physics 10*, 43-72. [III.4]

104. JAMMER, M. (1966). *The Conceptual Development of Quantum Mechanics*. McGraw-Hill, New York. [I.2]

105. JAMMER, M. (1974). *The Philosophy of Quantum Mechanics*. John Wiley & Sons, New York. [I.2]

106. JAUCH, J. M. (1964). The Problem of Measurement in Quantum Mechanics. *Helvetica Physica Acta 37*, 293-316. [IV.4.1]

107. JAUCH, J. M. (1968). *Foundations of Quantum Mechanics*. Addison-Wesley, Reading, Massachusetts. [I; II; III]

108. JAUCH, J. M., and BÁRON, G. (1972). Entropy, Information and Szilard's Paradox. *Helvetica Physica Acta 45*, 220-232. [III.4]

109. JAUCH, J. M., WIGNER, E. P., and YANASE, M. M. (1967). Some Comments Concerning Measurements in Quantum Mechanics. *Il Nuovo Cimento 48 B*, 144-151. [IV.4.1]

110. JOENSUU (1985). *Symposium on the Foundations of Modern Physics 1985*. P. Lahti and P. Mittelstaedt, eds., World Scientific, Singapore. [I.2]

111. JOENSUU (1987). *Symposium on the Foundations of Modern Physics 1987*. P. Lahti and P. Mittelstaedt, eds., World Scientific, Singapore. [I.2]

112. JOENSUU (1990). *Symposium on the Foundations of Modern Physics 1990*. P. Lahti and P. Mittelstaedt, eds., World Scientific, Singapore, 1991. [I.2]

113. JOOS, E., and ZEH, H. D. (1985). The Emergence of Classical Properties through Interaction with the Environment. *Zeitschrift für Physik B 59*, 223-243. [IV.4.3]

114. KAKAZU, K. (1990). Equivalence Classes and Generalized Coherent States in Quantum Measurement. Preprint, University of the Ryukyus. [V]

115. KAKAZU, K., and MATSUMOTO, S. (1990). Stability of Particle Trajectories and Generalized Coherent States. Preprint, University of the Ryukyus. [IV.4.4]

116. KALMBACH, G. (1986). *Measures and Hilbert Lattices*. World Scientific, Singapore. [I.2]

117. KHALFIN, L. A., and TSIRELSON, B. S. (1987). A Quantitative Criterion of the Applicability of the Classical Description within the Quantum Theory. In: Joensuu (1987), pp. 369-401. [IV.4]

118. KOCHEN, S. (1985). A New Interpretation of Quantum Mechanics. In: Joensuu (1987), pp. 151-169. [IV.3.2]

119. KOCHEN, S., and SPECKER, E. P. (1967). The Problem of Hidden Variables in Quantum Mechanics. *Journal of Mathematics and Mechanics 17*, 59-87. [I.2, II.2.2]

120. KRAUS, K. (1983). *States, Effects and Operations*. Springer-Verlag, Berlin. [I; II; III]

121. KUDAKA, S., MATSUMOTO, S., and KAKAZU, K. (1989). Generalized Coherent State Approach to Quantum Measurement. *Progress of Theoretical Physics 82*, 665-681. [IV.4.4]

122. KUDAKA, S., and MATSUMOTO, S. (1990). Quantum Measurement and Generalized Coherent State. In: Joensuu (1990), pp. 190-202. [IV.4.4]

123. LAHTI, P. J. (1985). Uncertainty, Complementarity and Commutativity. In: Cologne (1984), pp. 61-80. [III.7]

124. LAHTI, P. J. (1987). Complementarity and Uncertainty: Some Measurement-Theoretical and Information-Theoretical Aspects. In Joensuu (1987), pp. 182-208. [III.7]

125. LAHTI, P. J. (1990). Quantum Theory of Measurement and the Polar Decomposition of an Interaction. *International Journal of Theoretical Physics 29*, 339-350. [III.; IV.3.2]

126. LAHTI, P. J., and BUGAJSKI, S. (1985). Fundamental Principles of Quantum Theory II. *International Journal of Theoretical Physics 24*, 1051-1080. [III.7]

127. LEGGETT, A. (1980). Macroscopic Quantum Systems and the Quantum Theory of Measurement. *Supplement of the Progress of Theoretical Physics 69*, 80-100. [V]

128. LÉVY-LEBLOND, J. M. (1977). Towards A Proper Quantum Theory. In: *Quantum Mechanics, A Half Century Later*, J. Leite Lopes and M. Paty, eds., D. Reidel, Dordrecht, pp. 171-206. [V]

129. LINDBLAD, G. (1972). An Entropy Inequality for Quantum Measurements. *Communications in Mathematical Physics 28*, 245-249. [III.4.1]

130. LINDBLAD, G. (1973). Entropy, Information and Quantum Measurements. *Communications in Mathematical Physics 33*, 305-322. [III.4]

131. LINDBLAD, G. (1983). *Non-Equilibrium Entropy and Irreversibility*. Reidel, Dordrecht. [III.4]

132. LONDON, F., and BAUER, E. (1939). La Théorie de l'Observation en Méchanique Quantique. Hermann, Paris. English translation ("including a new paragraph by Professor Fritz Bauer") in Wheeler and Zurek (1983). [IV.1]

133. LOCKHART, C. M., and MISRA, B. (1986). Irreversibility and Measurement in Quantum Mechanics. *Physica 136A*, 47-76. [IV.5]

134. LUDWIG, G. (1961). Gelöste und ungelöste Probleme des Meßprozesses in der Quantenmechanik. In: *Werner Heisenberg und die Physik unserer Zeit*, F. Bopp, ed., Vieweg, Braunschweig, pp. 150-181. [IV.4.1]

135. LUDWIG, G. (1983a). *Foundations of Quantum Mechanics, Vol I*. Springer-Verlag, Berlin. [I; II; III; IV]

136. LUDWIG, G. (1983b). The Connection between the Objective Description of Macrosystems and Quantum Mechanics of "Many Par-

ticles". In: *Old and New Questions in Physics, Cosmology, Philosophy and Theoretical Biology*, A. van der Merve, ed., Plenum Press, New York, pp. 243-263. [IV.5.3]

137. LUDWIG, G. (1987). *An Axiomatic Basis for Quantum Mechanics, Vol. 2: Quantum Mechanics and Macrosystems*. Springer-Verlag, Berlin. [III.6; IV.5.3]

138. LÜDERS, G. (1951). Über die Zustandsänderung durch den Meßprozess. *Annalen der Physik 8*, 322-328. [III.3; III.4]

139. MACHIDA, S., and NAMIKI, M. (1980). Theory of Measurement – A Mechanism of Reduction of Wave Packet. *Progress of Theoretical Physics 63*, 1457-1473 (Part I) and 1833-1847 (Part II). [IV.5.1]

140. MACKEY, G. (1963). *The Mathematical Foundation of Quantum Mechanics*. W. A. Benjamin, New York. [I.2; III.5.2]

141. MARGENAU, H. (1936). Quantum Mechanical Description. *Physical Review 49*, 240-242. [I.2]

142. MARGENAU, H. (1963). Measurement and Quantum States. *Philosophy of Science 30*, 1-16. [IV.2]

143. MISRA, B., PRIGOGINE, I. and COURBAGE, M. (1979). Lyapunov Variable: Entropy and Measurement in Quantum Mechanics. *Proceedings of the National Academy of Sciences of the U. S. A. 76*, 4768-4772. (Reprinted in Wheeler and Zurek (1983).) [IV.4.1]

144. MITTELSTAEDT, P. (1976a). On the Applicability of the Probability Concept to Quantum Theory. In: *Foundations of Probability Theory, Statistical Inference and Statistical Theories of Science, Vol. III*. W. L. Harper and C. A. Hooker, eds., Reidel, Dordrecht, 1976, pp. 155-167. [II.2.3; IV.3.3]

145. MITTELSTAEDT, P. (1976b). *Philosophical Problems of Modern Physics*. Reidel, Dordrecht. [IV.3.3]

146. MITTELSTAEDT, P. (1978). *Quantum Logic*. Reidel, Dordrecht. [I.2]

147. MITTELSTAEDT, P. (1990). The Objectification in the Measuring Process and the Many Worlds Interpretation. In: Joensuu (1990), pp. 261-279. [III.2.4]

148. MUNICH (1981). *Quantum Optics, Experimental Gravity and Meas-urement Theory.* P. Meystre and M. O. Scully, eds., Plenum Press, New York, 1983. [I.2]

149. NELSON, E. (1985). Quantum Fluctuations, Princeton University Press, Princeton, New Jersey. [IV.2]

150. NEUMANN, J. VON (1932). *Mathematische Grundlagen der Quan-tenmechanik.* Springer-Verlag, Berlin. English translation: *Mathe-matical Foundations of Quantum Mechanics,* Princeton University Press, Princeton, 1955. [II, III]

151. NEW YORK (1986). *New Techniques and Ideas in Quantum Mea-surement Theory.* D. M. Greenberger,ed., Annals of the New York Academy of Sciences, 1987, Vol. 480. [I.2]

152. OCHS, W. (1977). On the Strong Law of Large Numbers in Quan-tum Probability Theory. *Journal of Philosophical Logic, 6,* 473-480. [III.2.4; IV.3.1]

153. OCHS, W. (1980). Gesetze der Großen Zahlen zur Auswertung quantenmechanischer Meßreihen. In: *Grundlagen der Quanten-theorie,* P. Mittelstaedt and J. Pfarr, eds., Bibliographisches Insti-tut, Mannheim, pp. 127-138. [III.2.4]

154. OMNÈS, R. (1988a). Logical Reformulation of Quantum Mechanics I. Foundations. *Journal of Statistical Physics 53,* 893-932. [IV.3.3]

155. OMNÈS, R. (1988b). Logical Reformulation of Quantum Mechan-ics II. Interference and the Einstein-Podolsky-Rosen Experiment. *Journal of Statistical Physics 53,* 933-955. [IV.3.3]

156. OMNÈS, R. (1988c). Logical Reformulation of Quantum Mechan-ics III. Classical Limit and Irreversibility. *Journal of Statistical Physics 53,* 957-975. [IV.3.3]

157. OZAWA, M. (1984). Quantum Measuring Processes of Continuous Observables. *Journal of Mathematical Physics 25,*79-87. [III]

158. OZAWA, M. (1986). On Information Gain by Quantum Meas-urements of Continuous Observables. *Journal of Mathematical Physics 27,* 759-763. [III.4.1]

159. PARTOVI, H. M. (1989). Irreversibility, Reduction and Entropy Increase in Quantum Measurements. *Physics Letters A137*, 445-450. [IV.4.3]

160. PAULI, W. (1933). *Die allgemeinen Prinzipien der Wellenmechanik*. In: *Handbuch der Physik*, H. Geiger and K. Scheel, eds., 2nd edition, Vol. 24, Springer-Verlag, Berlin, pp. 83-272. English Translation: *General Principles of Quantum Mechanics*, Springer-Verlag, Berlin, 1980. [I; III.3; IV.1]

161. PEARLE, P. M. (1984). Experimental Test of Dynamical State-Vector Reduction. *Physical Review D 29*, 235-240. [IV.4.2]

162. PEARLE, P. M. (1986). Stochastic Dynamical Reduction Theories and Superluminal Communication. *Physical Review D 33*, 2240-2252. [IV.4.2]

163. PEARLE, P. M. (1989). Combining Stochastic Dynamical State-Vector Reduction with Spontaneous Localization. *Physical Review A39*, 2277-2289. [IV.4.2]

164. PERCIVAL, I. C. (1961). Almost Periodicity and the Quantal H-Theorem. *Journal of Mathematical Physics 2*, 235-239. [IV.4.1]

165. PERES, A. (1980). Can we undo quantum measurements? *Physical Review D22*, 879-883. Reprinted in Wheeler and Zurek, 1983. [V]

166. PERES, A. (1987). Quantum Chaos and the Measurement Problem. In: *Quantum Measurement and Chaos*, E. R. Pike and S. Sarkar, eds., Plenum Press, New York, pp. 59-80. [V]

167. PERES, A., and ZUREK, W. H. (1982). Is Quantum Theory Universally Valid? *American Journal of Physics 50*, 807-810. [V]

168. PIRON, C. (1976). *Foundations of Quantum Physics*. Benjamin, Reading, Massachusetts. [III.5]

169. PRIMAS, H. (1983). *Chemistry, Quantum Mechanics and Reductionism*. Springer-Verlag, Berlin, 2nd. edition. [I.2, IV.5.2]

170. PRIMAS, H. (1990a). Induced Nonlinear Time Evolution of Open Quantum Objects. In: *Sixty-Two Years of Uncertainty: Historical, Philosophical, Physics Inquiries into the Foundations of Quantum Physics*, A. I. Miller, ed., Plenum Press, New York. [IV.4]

171. PRIMAS, H. (1990b). The Measurement Process in the Individual Interpretation of Quantum Mechanics. In: Rome (1989), pp. 49-68. [IV.5.2]

172. PRIMAS, H. (1990c). Necessary and Sufficient Conditions for an Individual Description of the Measurement Process. In: Joensuu (1990), pp. 332-346. [IV.5.2]

173. PRIMAS, H. (1990d). Mathematical and Philosophical Questions in the Theory of Open and Macroscopic Quantum Systems. In: *Sixty-Two Years of Uncertainty: Historical, Philosophical, Physics Inquiries into the Foundations of Quantum Physics*, A. I. Miller, ed., Plenum Press, New York. [V]

174. PROSPERI, G. M. (1971). Macroscopic Physics and the Problem of Measurement in Quantum Mechanics. In: *Foundations of Quantum Mechanics*, B. d'Espagnat, ed., Academic Press, New York, pp. 97-126. [IV.4.1]

175. PROSPERI, G. M. (1974). Models of the Measuring Process and Macro-Theories, in: *Foundations of Quantum Mechanics and Ordered Linear Spaces*, A. Hartkämper and H. Neumann, eds., Springer-Verlag, Berlin, pp. 163-198. [IV.4]

176. PRUGOVEČKI, E. (1986). *Stochastic Quantum Mechanics and Quantum Space Time*. D. Reidel Publishing Corporation, Dordrecht, 2nd edition. [I; II.1.1; III.7; IV.4.4]

177. QUADT, R. (1989). The Nonobjectivity of Past Events in Quantum Mechanics. *Foundations of Physics 19*, 1027-1035. [IV.3.3]

178. RAGGIO, G. (1982). States and Composite Systems in W^*-Algebraic Quantum Mechanics. Dissertation, ETH Zürich. [IV.5.2]

179. REED, M., and SIMON, B. (1972). *Methods of Modern Mathematical Physics*. I: Functional Analysis, Academic Press, New York. [II.1]

180. ROME (1989). *Quantum Theory without Reduction*. M. Cini and J. M. Lévy-Leblond, eds., Adam Hilger, Bristol and New York, 1990. [I]

159

181. ROSENFELD, L. (1965). The Measuring Process in Quantum Mechanics. *Supplement of the Progress of Theoretical Physics*, pp. 222-231. [IV.4.1]

182. SCHROECK, F. E., JR. (1981). A Model of a Quantum Mechanical Treatment of Measurement with a Physical Interpretation. *Journal of Mathematical Physics 22*, 2562-2572. [II.1.1; IV.4.4]

183. SCHRÖDINGER, E. (1935). Die gegenwärtige Situation in der Quantenmechanik. *Die Naturwissenschaften 23*, 807-812, 824-828, 844-849. [I.2]

184. SCHRÖDINGER, E. (1936). Probability Relations between Separated Systems. *Proceedings of the Cambridge Philosophical Society 32*, 446-452. [I.2]

185. SEWELL, G. L. (1986). *Quantum Theory of Collective Phenomena.* Clarendon Press, Oxford [IV.5.2]

186. SHERRY, T., and SUDARSHAN, E. C. G. (1978). Interactions Between Classical and Quantum Systems: A New Approach to Quantum Measurement I. *Physical Review D 18*, 4580-4589. [IV.5.1]

187. SHERRY, T., and SUDARSHAN, E. C. G. (1979). Interactions Between Classical and Quantum Systems: A New Approach to Quantum Measurement II. *Physical Review D 20*, 857-868. [IV.5.1]

188. SHIMONY, A. (1963). Role of the observer in quantum theory. *American Journal of Physics 31*, 755-773. [IV.1]

189. SHIMONY, A. (1974). Approximate Measurement in Quantum Mechanics, II. *Physical Review D9*, 2321-2323. [III.5.2]

190. SHIMONY, A. (1979). Proposed Neutron Interferometer Test of Some Nonlinear Variants of Wave Mechanics. *Physical Review A20*, 394-396. [IV.4.2]

191. STEIN, H., and SHIMONY, A. (1971). Limitations on Measurement. In: *Foundations of Quantum Mechanics*, B. d'Espagnat, ed., Academic Press, New York, pp. 56-76. [III.7.3]

192. STINESPRING, W. F. (1955). Positive Functions on C^*-Algebras. *Proceedings of the American Mathematical Society 6:I*, 211-216. [III.3.6]

193. TAKESAKI, M. (1979). *Theory of Operator Algebras.* Springer-Verlag, Berlin. [III.6.2]

194. THIRRING, W. (1980). *A Course in Mathematical Physics, Vol. 4. Quantum Mechanics of Large Systems.* Springer-Verlag, New York, Wien. [III.4]

195. TOKYO (1983). *Proceedings of the International Symposium on the Foundations of Quantum Mechanics in the Light of new Technology.* S. Kamefuchi, H. Ezawa, Y. Murayama, M. Namiki, S. Nomura, Y. Ohnuki and T. Yajima, eds., Physical Society of Japan, 1984. [I.2]

196. TOKYO (1986). *Proceedings of the 2nd International Symposium on the Foundations of Quantum Mechanics in the Light of new Technology.* M. Namiki, Y. Ohnuki, Y. Murayama and S. Nomura, eds., Physical Society of Japan, 1987. [I.2]

197. TOKYO (1989). *Proceedings of the 3rd International Symposium on the Foundations of Quantum Mechanics in the Light of new Technology.* S. Kobayashi, H. Ezawa, Y. Murayama and S. Nomura, eds., Physical Society of Japan, 1990. [I.2]

198. TZARA, C. (1987). Fuzzy Measurements in Quantum Mechanics and the Representation of Macroscopic Motions. *Il Nuovo Cimento 98 B*, 131-143. [IV.4.4]

199. TZARA, C. (1988). Emergence of a Classical Motion from a Quantum State: A Further Test of a Theory of Fuzzy Measurements. *Physics Letters A 127*, 247-250. [IV.4.4]

200. VAN FRAASSEN, B. C. (1979). Foundations of Probability: A Modal Frequency Interpretation. In: *Problems in the Foundations of Physics*, G. Toraldo di Francia, ed., North-Holland Publishing Corporation, Amsterdam, pp. 344-394. [III.2.4; IV.3.3]

201. VAN FRAASSEN, B. C. (1981). A Modal Interpretation of Quantum Mechanics. In: *Current Issues in Quantum Logic*, E. Beltrametti and Bas C. van Fraassen, eds., Plenum Press, New York, pp. 229-258. [IV.3.3]

202. VAN FRAASSEN, B. C. (1990). The Modal Interpretation of Quantum Mechanics. in: Joensuu (1990), pp. 440-460. [IV.3.3]

203. VARADARAJAN, V. S. (1985). *Geometry of Quantum Theory.* Second Edition, Springer-Verlag, New York. [I]

204. VIENNA (1987). *Matter Wave Interferometry.* G. Badurek, H. Rauch and A. Zeilinger, eds., North Holland, Amsterdam, 1988. [I.2]

205. WALLS, D. F., COLLET, M. J., and MILBURN, G. J. (1985). Analysis of a Quantum Measurement. *Physical Review D32*, 3208-3215. [IV.4.3]

206. WAN, K. K. (1980). Superselection Rules, Quantum Measurement and the Schrödinger's Cat. *Canadian Journal of Physics 58*, 976-982. [IV.4.1]

207. WEIDLICH, W. (1967). Problems of the Quantum Theory of Measurement. *Zeitschrift für Physik 205*, 199-220. [IV.4.1]

208. WEINBERG, S. (1989). Testing Quantum Mechanics. *Annals of Physics 194*, 336-386. [IV.4.2]

209. WHEELER, J. A. (1957). Assessment of Everett's "relative" State Formulation of Quantum Theory. *Reviews of Modern Physics 29*, No. 3, 463-465. [IV.3.1]

210. WHEELER, J. A., and ZUREK, W. H. (1983). *Quantum Theory and Measurement.* Princeton University Press, Princeton. [I-IV]

211. WHITTEN-WOLFE, B., and EMCH, G. G. (1976). A Mechanical Quantum Measuring Process. *Helvetica Physica Acta 49*, 45-55. [IV.5.2]

212. WIGNER, E. P. (1952). Die Messung Quantenmechanischer Operatoren. *Zeitschrift für Physik 133*, 101-108. [III.7]

213. WIGNER, E. P. (1961). Remarks on the Mind-Body Question. In: *The Scientist Speculates*, I. J. Good, ed., W. Heinemann, London, 1961. [IV.1]

214. WIGNER, E. P. (1963). The Problem of Measurement. *American Journal of Physics 31*, 6-15. [IV.1]

215. WIGNER, E. P. (1983). Review of the Quantum Mechanical Measurement Problem. In: Munich (1981), pp. 43-63. [IV.1]

216. WÖLFEL, J. (1987). *Über Folgen quantenmechanischer Messungen und deren informationstheoretische Behandlung.* Dissertation, Cologne. [III.4.3]

217. ZAORAL, W. (1990). Towards a Derivation of a Nonlinear Stochastic Schrödinger Equation for the Measurement Process from Algebraic Quantum Mechanics. In: Joensuu (1990), pp. 479-486. [IV.5.2]

218. ZEH, D. (1970). On the Interpretation of Measurement in Quantum Theory. *Foundations of Physics 1*, 69-76 (1970). [IV.1; IV.4.3]

219. ZUREK, W. H. (1981). Pointer Basis of Quantum Apparatus: Into What Mixture does the Wave Packet Collapse? *Physical Review D24*, 1516-1525. [IV.4.3]

220. ZUREK, W. H. (1982). Environment-induced Superselection Rules. *Physical Review D 26*, 1862-1880. [IV.4.3]

221. ZUREK, W. H. (1984). Pointer Basis and Inhibition of Quantum Tunneling by Environment-Induced Superselection. In: Tokyo (1983), pp. 181-189. [IV.4.3]

List of frequently appearing notations

c_X 45

E 9, 10

E^A 9, 10

E_T 10

$E_T(X)$ 10

$\mathcal{E}(\mathcal{H}_S)$ 12

γ_i 39

φ_{a_i} 41

$\mathcal{I}_L$ 41

$\mathcal{I}_\mathcal{M}$ 35

$\mathcal{I}_{vN}$ 32

$\mathcal{I}_U$ 40

$\mathcal{L}(\mathcal{H})$ 8

$\mathcal{L}(\mathcal{H}_S)^+$ 10

$\mathcal{M}$ 34

$\mathcal{M}_U$ 36

$\mathcal{M}_U^m$ 39

N_i^2 38, 63

$P_A^\mathcal{R}$ 44

$p_\varphi(a_i)$ 38

$\mathcal{R}$ 44

$\mathcal{R}_A\big(V(T \otimes T_A)\big)$ 15, 32

$\mathcal{R}_S\big(V(T \otimes T_A)\big)$ 15, 32

$T_{A,i}$ 76

T_{A,Ω_A} 75

T_Ω 63

$T_{\Omega,i}$ 63

T_U 77

$T_{A,U}$ 77

$\mathcal{T}(\mathcal{H}_S)_1^+$ 10

$tr\,[TE(X)]$ 10

$\mathcal{U}_t(T_0)$ 16

U_L 41

$\mathcal{V}_t$ 16

$|\varphi_i\rangle\langle\varphi_j|$ 15

$\langle \mathcal{H}_A, P_A, T_A, V, f \rangle$ 32, 34

$\langle \mathcal{H}_A, P_A, \Phi, U \rangle$ 36

$\langle \mathcal{H}_A, A_A, \Phi, U \rangle$ 39

$\langle \mathcal{H}_A, A_A, \Phi, U_L \rangle$ 41

$(\Omega, \mathcal{F})$ 9

$(\Omega_A, \mathcal{F}_A)$ 32

$(\mathfrak{R}, \mathcal{B}(\mathfrak{R}))$ 9

$T_1 \cong_\mathcal{O} T_2$ 20

$T_1 \cong T_2$ 20

Index

165